Practice for Book S2

Contents

CAMBRIDGE
UNIVERSITY PRESS

PUBLISHED BY THE PRESS SYNDICATE OF THE UNIVERSITY OF CAMBRIDGE
The Pitt Building, Trumpington Street, Cambridge, United Kingdom

CAMBRIDGE UNIVERSITY PRESS
The Edinburgh Building, Cambridge CB2 2RU, UK
40 West 20th Street, New York, NY 10011-4211, USA
477 Williamstown Road, Port Melbourne, VIC 3207, Australia
Ruiz de Alarcón 13, 28014 Madrid, Spain
Dock House, The Waterfront, Cape Town 8001, South Africa
http://www.cambridge.org

Printed in the United Kingdom at the University Press, Cambridge
Typeface Minion   *System* QuarkXPress®

A catalogue record for this book is available from the British Library

ISBN 0 521 79868 X   paperback

Typesetting and technical illustrations by The School Mathematics Project
Illustrations on pages 8, 64 and 65 by Chris Evans
Crown copyright material on page 28 is reproduced with the permission of the Controller of
HMSO and the Queen's Printer for Scotland.
Cover image © Image Bank/Antonio Rosario
Cover design by Angela Ashton

① Into the bath

1 Ffion is timing a hotel lift as it moves between floors.

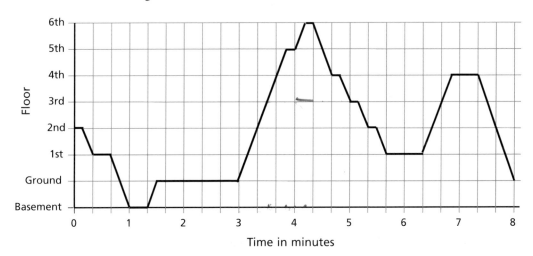

When she starts timing, at 0 minutes, the lift is at the 2nd floor.

(a) Write down what the lift does in the first minute.

(b) What do you think happens at 3 minutes?

(c) How long is the lift at the 5th floor?

(d) Sam is on the 3rd floor.
   At 3 minutes he pushes the button for the lift.
   How long does he have to wait for the lift to arrive?

(e) During the 8 minutes shown in the graph, Gill takes the lift up to
   the 4th floor.
   At what floor do you think she gets in the lift?

(f) For how long during these 8 minutes is the lift above the level of
   the 3rd floor?

2 Write a story for this graph.

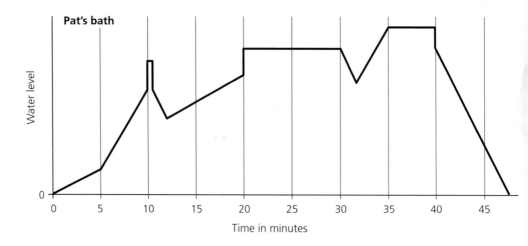

3 This graph shows the number of people in a shopping centre one day near Christmas.

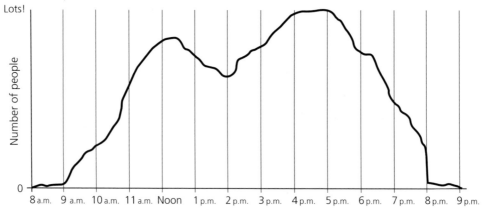

(a) At what time do you think the shopping centre opened?

(b) What time do you think it closed?

(c) At roughly what time were there most people in the centre?

(d) Were there more people in the centre at 3 p.m. or 4 p.m.?

(e) At 11 a.m., were more people coming in to the centre or leaving it?

(f) At 1 p.m., were more people coming in or leaving?

② Ratio

Section A

1   Most gold used for jewellery is a mixture of
    pure gold and other metals.
    '18 carat' gold is made by mixing 3 parts of
    pure gold with 1 part of other metals by weight.

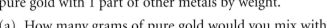

(a) How many grams of pure gold would you mix with

    (i) 5 g of other metals   (ii) 12 g of other metals

(b) How many grams of other metals would you mix with

    (i) 30 g of gold   (ii) 42 g of gold

2   Here is a recipe for 9 carat gold.

    Copy and complete the table.

Mix 3 parts of pure gold
with 5 parts of other metals.

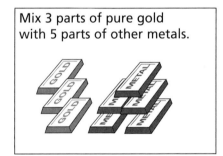

Pure gold (g)	Other metals (g)
6	
12	
15	
	15
	30
	75

3   Here is the recipe for 14 carat gold.

    Copy the table and fill it in.

Mix 7 parts of pure gold
with 5 parts of other metals.

Pure gold (g)	Other metals (g)
7	
21	
	20
35	
	35
	90

Section B

1 These brooches are made from triangular and square shapes.

What is the ratio of triangles to squares for each design?

(a)   (b)   (c)   (d)   (e)

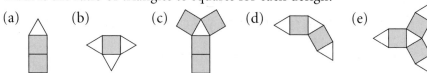

2 (a) Which of these sets of squares and
triangles can be made into designs like this
with no squares or triangles left over?

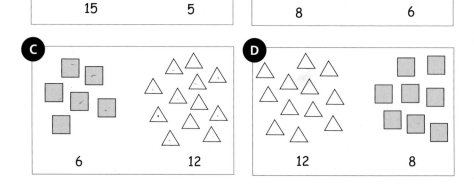

(b) Match each of these designs to one of the sets above and write down
the ratio of triangles to squares.

(i)   (ii)   (iii)   (iv)   (v)

Section E

1 Amy made a necklace using 24 red beads and 18 green beads.

   (a) Write the ratio of red beads to green beads in its simplest form.

   (b) Write the ratio of green beads to red beads in its simplest form.

2 Write each of these ratios in its simplest form.

   (a) 4:6   (b) 12:4   (c) 15:25   (d) 15:6   (e) 21:18

3 Complete these ratios so that all the ratios in each group are equal:

   (a) 2:3   ■:9   10:■   (b) 4:1   ■:3   8:■

   (c) 6:2   ■:1   18:■   (d) 4:10   2:■   ■:15

   (e) 30:40   ■:8   3:■   (f) 7:3   21:■   ■:12

4 Match each ratio from group A with an equal ratio from group B.

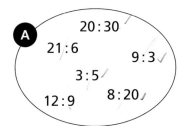
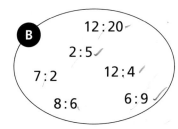

5 A compost for growing cactus plants is made by mixing loam and sand
   in the ratio of 3:1.

   (a) How many bags of loam go with 3 bags of sand?

   (b) How many bags of sand go with 12 bags of loam?

   (c) How many bags of loam and how many bags of sand are needed to
        make 20 bags of cactus compost?

6 Purple paint is made by mixing blue and red in the ratio of 3:2.

   (a) How many litres of red paint go with 6 litres of blue paint?

   (b) How many litres of blue paint go with 8 litres of red paint?

   (c) How many litres of blue paint and how many litres of red paint
        are needed to make 25 litres of purple paint?

Section F

1 Two friends Mike and Roxanne share a pile of sweets in the ratio of $2:1$.
They have 36 sweets in total.

   (a) How many sweets does each get?

   (b) What fraction of the sweets does Roxanne get?

2 Jenny and Helen share £50 in the ratio of $1:4$.

   (a) How much do they each get?

   (b) What fraction of the money does Jenny get?

   (c) What fraction of the money does Helen get?

3 Class 8C were going to raise some money for charity.
They decided to share it between two charities in the ratio of $2:3$.

   Write down how much each charity would receive if they raised

   (a) £10   (b) £25   (c) £80   (d) £42.50

4 Paul and Anna were being paid £30 to do some gardening.
They decided to share the payment according to
how many hours they worked.

   (a) How much was Anna's share if
she worked for 2 hours and
Paul worked for 4 hours?

   (b) What was Paul's share if he
worked for 2 hours and
Anna worked for 3 hours?

   (c) How much did they each get if Paul worked for 5 hours
and Anna worked for 3 hours?

5 Andrew and David were candidates in a school election.
Andrew received more votes than David.

   The votes were cast for the two candidates in the ratio $5:4$.

   135 children voted.

   How many more votes did Andrew receive than David?

6 Claire and Emma shared a packet of 21 sweets so that Claire had twice as many sweets as Emma.

How many did they each get?

7 Gert, Hans and Ivy share 24 sweets in the ratio $1:3:2$.

How many sweets did they each get?

8 (a) Share £20 in the ratio $1:1:3$.   (b) Share £35 in the ratio $2:1:4$.

   (c) Share £100 in the ratio $2:3:5$.   (d) Share £99 in the ratio $1:4:6$.

Section G

1 Find the missing numbers.

  (a) $2:3 = 4:\blacksquare$ (b) $1:5 = \blacksquare:25$ (c) $4:7 = 12:\blacksquare$

2 For each pair of colours, work out which is the darker.

(a)
Pastel pink	Mix red and white in the ratio $2:5$.
Pearly pink	Mix red and white in the ratio $3:10$.

(b)
Gun grey	Mix black and white in the ratio $3:4$.
Slate grey	Mix black and white in the ratio $4:7$.

(c)
Boy blue	Mix blue and white in the ratio $5:8$.
Sky blue	Mix blue and white in the ratio $2:3$.

3 A recipe for Lime Fizz mixes lime juice and soda in the ratio $5:12$.
A recipe for Tangy Lime mixes lime juice and soda in the ratio $4:9$.

(a) Write each ratio in the form $1:\blacksquare$.

(b) Which drink uses the greater proportion of lime juice?

4 A recipe for flaky pastry uses 175 grams of butter with 225 grams of flour.
A recipe for choux pastry uses 50 grams of butter with 65 grams of flour.

Which type of pastry uses the greater proportion of butter?

③ Starting equations

Section A

1   Write the equation shown by the picture balance.
Solve the equation and check your solution.

(a)

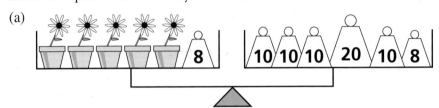

(b)

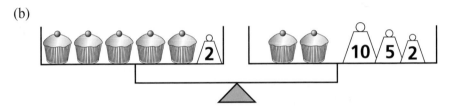

(c)

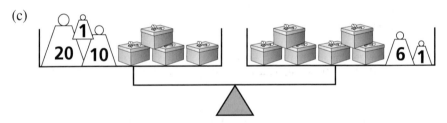

(d)

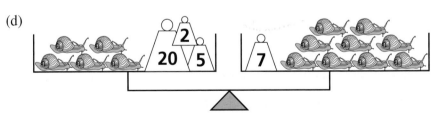

(e)

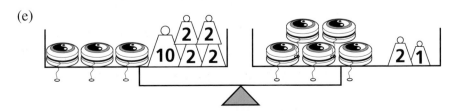

Section B

1 Solve these equations and check each solution.

(a) $6n + 1 = 2n + 9$ (b) $2w + 16 = 4w + 4$ (c) $5h + 3 = 3h + 5$

(d) $7 + 3t = 19$ (e) $4x + 7 = 23$ (f) $5p + 4 = 12 + 3p$

(g) $6d + 3 = 3d + 27$ (h) $5m + 2 = 3m + 8$ (i) $6k = 2k + 28$

(j) $9m + 1 = 3m + 13$ (k) $5p + 4 = 6p$ (l) $5y + 4 = 1 + 6y$

2 Solve these equations and check your answers.

(a) $2t + 5 = 23$ (b) $14 + t = 8t$ (c) $9 + 2t = 3 + 4t$

(d) $t + 15 = 5t + 3$ (e) $10 + 3t = t + 21$ (f) $3t + 1 = t + 4$

(g) $7t + 3 = 43 + 3t$ (h) $10t + 1 = 5t + 7$ (i) $5t + 1 = 4t + 1$

Section C

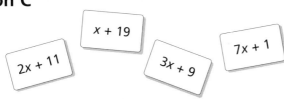

1 Jay picks two of these cards and uses them to make this equation.

$$2x + 11 = 7x + 1$$

Solve this equation to find the value of x which makes it true.

2 With another pair of cards, she makes the equation

$$x + 19 = 3x + 9$$

Find the value of x which makes it true.

3 Find some more different equations using these cards.
Solve each equation.

Section D

1   Billy and Gina both start with the same number.

I multiply by 10
and add 3.

I multiply by 6
and add 27.

Amazingly, they get the same result.

Form an equation and work out what number they were thinking of.

2   Abda and Leena both start with the same number.

I multiply by 4
and add 80.

I multiply by 20
and add 48.

Their results are the same.

What number did they start with?

3   Marisa and Fabio both start with the same number.

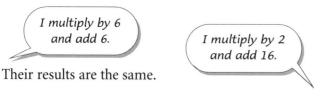

I multiply by 6
and add 6.

I multiply by 2
and add 16.

Their results are the same.

What number did they start with?

*4   Anna and Sophie both start with the same number.
     Anna adds 5 then multiplies the result by 3.
     Sophie adds 6 then multiplies the result by 2.
     Their answers are the same.

What number did they start with?

④ Cuboids

Section A

1   This cuboid is made
    using 30 cubes.

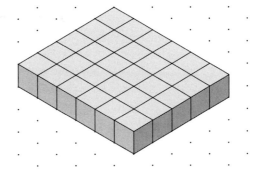

On triangular dotty paper draw a
different cuboid that is made of 30 cubes.

2   Sketch as many different cuboids as you can that are made of 32 cubes.

3   How many cubes are there in each of these cuboids?

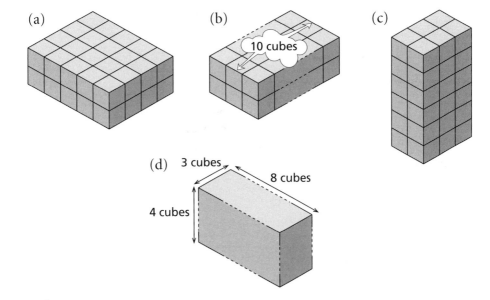

(a)

(b)   10 cubes

(c)

(d)   3 cubes

8 cubes

4 cubes

Section B

1 Work out the volume of each of these cuboids.

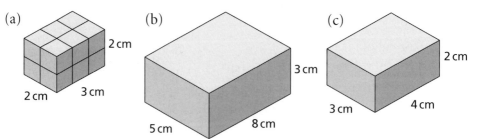

(a) 2 cm, 2 cm, 3 cm

(b) 3 cm, 5 cm, 8 cm

(c) 2 cm, 3 cm, 4 cm

2 Work out the volume of each of these cuboids.

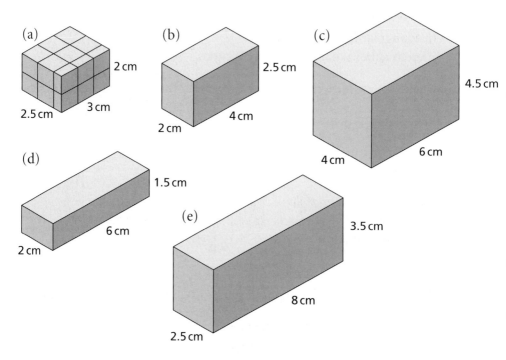

(a) 2 cm, 2.5 cm, 3 cm

(b) 2.5 cm, 2 cm, 4 cm

(c) 4.5 cm, 4 cm, 6 cm

(d) 1.5 cm, 6 cm, 2 cm

(e) 3.5 cm, 8 cm, 2.5 cm

3 Calculate the volume of each of these cuboids.

    (a) 6.5 cm by 8 cm by 4 cm   (b) 7.5 cm by 9 cm by 10 cm

    (c) 16 cm by 12.5 cm by 8.5 cm   (d) 4.5 cm by 12 cm by 3.5 cm

4 The volume of each of these cuboids is 50 cm³.
Find the missing lengths.
(All dimensions are in cm.)

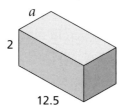 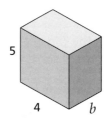 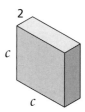

Section C

1 This shape is made from two cuboids.

(a) Find its volume.

(b) Find its surface area.

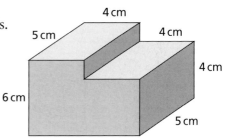

Section D

1 A normal coffee mug holds about 250 ml.
How many mugs like this could you fill from an urn containing
30 litres of coffee?

2 Write these quantities as decimals of a litre.

(a) 570 ml   (b) 350 ml   (c) 7 ml   (d) 32 c.c.

3 Write these quantities in order, smallest first.

410 cm³   48 ml   3.5 litres   0.4 litre   305 ml

4 Write these quantities in millilitres.

(a) $\frac{1}{5}$ litre   (b) $\frac{7}{10}$ litre   (c) $\frac{3}{4}$ litre   (d) $\frac{2}{5}$ litre

5 Graphs, charts and tables

Section A

1   This chart shows the sorts of jobs people had in nine countries in the European Union in 1996.

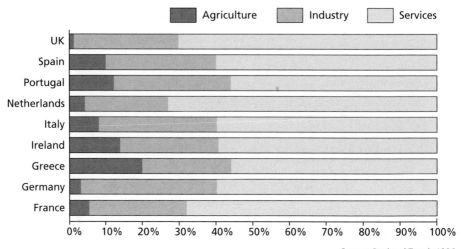

Source: Regional Trends 1996

(a) Which country had the smallest percentage of people employed in agriculture?

(b) Which country had most employed in agriculture?

(c) Which countries had the smallest percentage of people employed in services?

(d) Which country had most in services?

(e) Which country had most in industry?

(f) Which countries had 30% or more employed in industry?

(g) Which countries had 60% or less employed in services?

2 In a survey, year 7 and year 11 pupils were asked how they got to school. This chart shows the results as percentages.

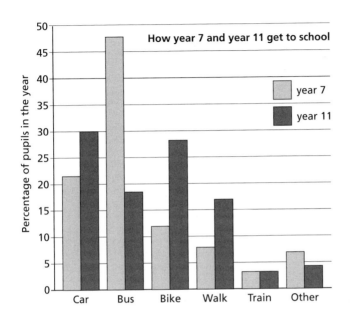

(a) What was the most common way for year 7 pupils to get to school?

(b) What was the most common way for year 11?

(c) What percentage of year 11 pupils came by car?

(d) Roughly what percentage of year 7 pupils walked?

(e) Did more year 7 pupils come by car or by bike?

(f) For year 11, which was more common, to come by bike or by bus?

(g) In year 8, 20% of the pupils came by bike.
    Is this more or less than year 7?

(h) In year 8, $\frac{1}{4}$ of the pupils came by car.
    Is this more or less than year 7?

Section B

1 The graph below shows the temperatures in °C at noon in Sydney, London and Madrid.
It covers 15 days in May.

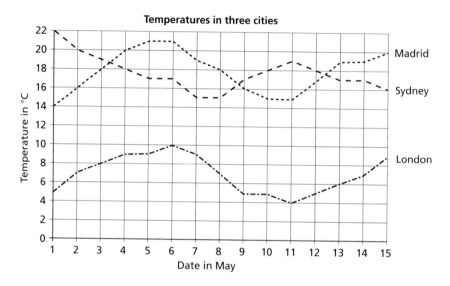

(a) Which city had the lowest noon temperature?
What was it?

(b) Which city had the highest temperature?
What was it?

(c) The temperature in London went up, then fell to its lowest on 11 May, and then went up again.
Describe what happened to the temperature in Sydney.

(d) On which dates was it hotter at noon in Madrid than in Sydney?

(e) On what date was there the greatest difference between the noon temperature in Madrid and in Sydney? What was the difference?

(f) What is the highest temperature in London shown on the graph?

(g) On how many days was the noon temperature in London 8°C or above?

Section C

1 In a school, pupils in years 7, 9 and 11 were asked how many minutes
   they spent on homework the previous night.
   The results for the three year groups are shown below.

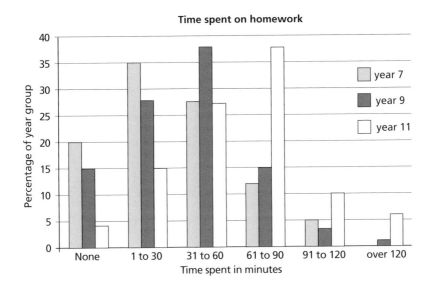

Time spent on homework

(a) In which year did the greatest percentage of pupils do no homework?

(b) In year 9, did more pupils do 31 to 60 minutes of homework
    or 61 to 90 minutes?

(c) What percentage of year 7 did 91 to 120 minutes of homework?

(d) What percentage of year 11 did 91 to 120 minutes of homework?

(e) About what percentage of year 11 did over 90 minutes of homework?

(f) About what percentage of year 9 did 30 minutes or less
    of homework?

(g) Who does most homework, year 7 or year 11?
    Explain how you can tell from the graph.

2   Represent each of these sets of data graphically in a suitable way.

(a) These are the weights in grams of 50 fish caught in a pond.

17	56	34	18	19	54	8	45	43	21
33	38	32	21	41	36	32	28	29	30
42	51	18	55	12	10	40	39	27	44
17	54	34	24	48	59	32	50	42	43
9	43	38	35	55	12	42	48	49	41

(b) This table shows the rainfall in Miami one year.
It shows how much rain fell each month in millimetres.

Jan	Feb	Mar	Apr	May	Jun	Jul	Aug	Sep	Oct	Nov	Dec
58	52	56	65	185	195	137	175	215	220	95	45

(c) This table shows the amount of homework done by pupils in year 8 and year 10 one night.
The figures shown are the percentages of each year group.

Time (min)	Year 8	Year 10
None	18	12
1 to 30	37	30
31 to 60	33	42
61 to 90	7	12
91 to 120	3	4
over 120	2	0

⑥ Fractions, decimals and percentages

Sections A and B

1 Evaluate these.

(a) $\frac{2}{5}$ of 25   (b) $\frac{2}{3}$ of 12   (c) $\frac{5}{7}$ of 84

(d) $\frac{1}{5}$ of 60   (e) $\frac{3}{10}$ of 120   (f) $\frac{5}{12}$ of 60

2 Some pupils in year 8 took part in a sponsored silence. This table shows the number from each tutor group.

Tutor group	Number of pupils
8G	12
8Y	10
8R	13
8L	8
8W	5

(a) How many pupils took part in the sponsored silence?

(b) What fraction of these pupils were from 8G?

(c) What fraction of these pupils were from 8L?

3 This table shows the number of pupils in each year group at High Hill School.

Year	Number of pupils
7	180
8	215
9	225
10	192
11	208

(a) $\frac{3}{4}$ of year 7 took part in a fun run. How many pupils was that?

(b) $\frac{4}{5}$ of year 9 did a sponsored silence. How many pupils was that?

(c) 52 of pupils in year 11 took part in a fun run. What fraction of year 11 took part in the fun run?

(d) How many pupils are there in the school in total?

(e) 765 pupils took part in one of the sponsored events. What fraction of the school is this ?

4 Write these fractions as decimals, rounding to two decimal places where necessary.

(a) $\frac{4}{5}$ (b) $\frac{1}{4}$ (c) $\frac{9}{10}$

(d) $\frac{5}{8}$ (e) $\frac{5}{6}$ (f) $\frac{6}{9}$

5 Write these percentages as decimals.

(a) 65%   (b) 40%   (c) 7%

(d) 75%   (e) 8%   (f) 12%

(g) 15%   (h) 91%   (i) 6%

6 Write these decimals as percentages.

(a) 0.71   (b) 0.8   (c) 0.09

7 Write each of these fractions as a percentage.

(a) $\frac{3}{4}$ (b) $\frac{4}{5}$ (c) $\frac{7}{10}$

(d) $\frac{3}{20}$ (e) $\frac{12}{25}$ (f) $\frac{11}{20}$

8 Write each of these fractions as a percentage, correct to the nearest 1%.

(a) $\frac{1}{3}$ (b) $\frac{1}{6}$ (c) $\frac{6}{13}$

(d) $\frac{2}{11}$ (e) $\frac{3}{17}$ (f) $\frac{5}{52}$

9 Calculate these.

(a) 25% of 56 kg   (b) 42% of £290   (c) 37% of 186 kg

(d) 3% of 176 g   (e) 18% of £68   (f) 99% of 358 cm

10 Calculate these (correct to two decimal places).

(a) 27% of £25.25   (b) 65% of 7.6 metres

(c) 32% of 134.9 kg   (d) 7% of 52.3 g

11 24 staff out of 58 work part time.

What percentage of staff are part-timers?

12 This table shows the number of people in each age group who took part in a sponsored cycle ride.

Age group	Number of people
10–19	62
20–25	144
26–30	35
31–40	24
41–50	16
51+	9
Total	**290**

(a) Decide if each statement is true or false.

    A About 50% of the cyclists were aged 20 to 25.

    B About 25% of the cyclists were aged 31 to 40.

    C Less than $\frac{1}{4}$ of the cyclists were aged 10 to 19.

    D Nearly $\frac{1}{8}$ of the cyclists were aged 26 to 30.

(b) What percentage of the cyclists were aged 10–19?

(c) What percentage were aged over 50?

13 This pie chart shows the sources of some money collected by a charity in 1998. They collected £6 228 000 altogether.

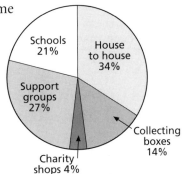

(a) What percentage was collected from charity shops?

(b) How much money was collected from charity shops?

(c) How much money was collected in house-to-house visits?

Section C

1   This table shows some population data for 1988.

	Population	Less than 15 years	Over 65 years
England	47 536 500	19%	16%
Wales	2 857 000	19%	16%
Scotland	5 094 000	19%	15%
Northern Ireland	1 578 100	25%	12%

For each of the four areas, to the nearest thousand,

(a) work out the number of people under 15

(b) work out the number of people over 65

2   This chart shows some road accident statistics for Great Britain from 1930 to 1990.

Year	Fatality				Total number killed
	Pedestrians	Pedal-cyclists	Motor-cyclists	Car drivers	
1930	3722	887	1832	864	7305
1950	2251	805	1129	827	5012
1970	2925	373	761	3440	7499
1990	1694	256	659	2608	5217

(a) What percentage of those killed in 1950 were motorcyclists?

(b) Copy and complete this table to show the percentages for each year.

Year	Pedestrians	Pedal-cyclists	Motorcyclists	Car drivers
1930	51%	12%		
1950	45%			
1970	39%			
1990				

(c) Draw pie charts, one for each year, to show the fatalities.

(d) Use your charts to describe the changes in the fatalities over the years.

7 Using rules

Section B

Pat's Painting Weekends organise group holidays where you can learn to paint. Pat uses rules to work out quantities of materials to take for the number in the group.

1 These are the rules Pat uses for the number of easels and brushes.

$$e = n + 2$$

$$b = 4n + 5$$

  n stands for the number of people
  e stands for the number of easels to take
  b stands for the number of brushes to take

(a) Write each rule as a sentence.

(b) Five people go on a painting weekend.
    How many easels and brushes will Pat take for this group?

(c) Work out *e* when $n = 10$.

(d) Work out *b* when $n = 6$.

(e) Pat takes 20 easels for a group.
    How many people are in the group?

2 This is the rule for watercolour sets.

    Take one watercolour set for every four people

(a) Which of these rules is correct for the number of watercolour sets?

$$w = 4n$$ $$w = n - 4$$ $$w = 4 + n$$ $$w = \frac{n}{4}$$

  n stands for the number of people
  w stands for the number of watercolour sets

(b) How many watercolour sets would be taken for 32 people?

(c) Pat takes 5 watercolour sets for a group.
    How many people are in the group?

(d) If $n = 16$, what is *w*?

3 Write each of these rules in shorthand. Choose your own letters for your rules, but say what each letter stands for.

(a) *Number of tubes of paint = (number of people × 6) + 4*

(b) *Take three sketch pads for each person.*

(c) *To find the number of picnic tables to take, divide the number of people by four, then add one extra.*

Section C

1 Mr Brown orders materials for a new geography department. These are the rules he uses. (*c* stands for the number of children.)

$$a = \frac{c}{2}$$
a is the number of atlases

$$g = \frac{c}{20} + 2$$
g is the number of globes

$$m = \frac{c - 10}{2}$$
m is the number of maps

Work out how many of each item he has to buy for

(a) 100 children   (b) 400 children   (c) 600 children

2 Use the rule $h = 2(g + 1)$ to work out h when

(a) $g = 2$ (b) $g = 5$ (c) $g = 20$

3 Use the rule $f = 5(d - 10)$ to work out f when

(a) $d = 12$ (b) $d = 10$ (c) $d = 50$

4 Use the rule $k = n^2$ to find what k is when

(a) $n = 4$ (b) $n = 1$ (c) $n = 12$

5 Use the rule $y = x^2 - 4$ to find what y is when

(a) $x = 6$ (b) $x = 10$ (c) $x = 4$

6 Work these out.

(a) $s^2 - 4$ when $s = 3$ (b) $t^2 + 10$ when $t = 0$

(c) $3(4 + x)$ when $x = 2$ (d) $\frac{y - 8}{4}$ when $y = 12$

(e) $6 + \frac{a}{2}$ when $a = 10$ (f) $4(t - 1)$ when $t = 10$

7 Here is a rule. $c = 3(a - 2b)$ Work out c when

(a) $a = 10$ and $b = 2$ (b) $a = 5$ and $b = 2$ (c) $a = 8$ and $b = 4$

Mixed questions 1

1 Share £140 in the ratio 2:5.

2 In 2000, 38% of the 25 000 000 households in the UK had a home computer. How many households had a computer?

3 There are 240 pupils in Year 8 at Tinton School.

(a) If $\frac{5}{8}$ of these pupils bring a packed lunch, how many is that?

(b) 180 of the Year 8 pupils walk to school. What percentage is this?

4 A liquid fertiliser has to be mixed with water in the ratio 1:25.

How much water should be used with 20 ml of fertiliser?

5 Here is a rule for the number of kebabs Pete prepares for a barbecue.

$k = 3n + 5$

where n is the number of people and k is the number of kebabs.

(a) How many kebabs will he prepare for 7 people?

(b) One day he prepared 50 kebabs.
How many people was he expecting at the barbecue?

6 Maria multiplied a number by 8 then added 21 and her answer was 11 times her original number.
Form an equation and solve it to find her original number.

7 (a) Work out the volume of this cuboid.

(b) Work out the surface area of this cuboid.

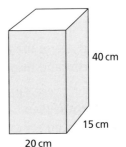

40 cm

15 cm

20 cm

8 (a) If $c = \dfrac{d-2}{3}$ find c when $d = 50$.

(b) If $y = 5(z + 3)$ find y when $z = 8$.

(c) If $r = n^2 - 2$ find r when $n = 7$.

9 Here are two recipes for elderflower cordial.

    A Use 3 parts of elderflower concentrate with 5 parts of sugar syrup.

    B Mix 5 parts of elderflower concentrate with 8 parts of sugar syrup.

Which cordial will have the stronger elderflower flavour?

10 Solve these equations.

    (a) $7x + 3 = 66$ (b) $5t = 16 + 3t$ (c) $7g + 2 = 20 + 4g$

11 Find the missing measurements of these cuboids.

(a) 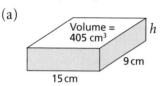   (b)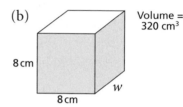

12 Write the ratio $120:60:45$ in its simplest form.

13 This graph shows, in millions, the births and deaths in the United Kingdom.

© Crown copyright 2002   Source: National Statistics

    (a) Approximately how many people were born in 2000?

    (b) When in the last century were the number of births and number of deaths about the same?

    (c) In which year, approximately, was there the largest number of births last century?

    (d) The number of births was more than the number of deaths for nearly every year last century.
        What does this tell us about the size of the population?

    (e) What does the graph suggest may happen to the size of the population after about 2025?

Decimals and area

Section A

1 Work out the areas of these rectangles.

(a) 1.5 m by 1.6 m (b) 1.4 m by 2.5 m (c) 3.8 m by 1.5 m

(d) 3.1 m by 4.1 m (e) 3.5 m by 7.2 m (f) 8.2 m by 0.5 m

2 Find the area of each square. Write about any patterns you notice.
   Can you explain them?

1.5 cm
1.5 cm

2.5 cm
2.5 cm

3.5 cm
3.5 cm

4.5 cm
4.5 cm

Section B

1 These are full-size designs for jewellery letters.
   Measure and find the area of each one.
   Draw sketches to show how you found your answers.

⑨ Finding rules

Sections A and B

1 These tiles are arranged to form T-shapes.

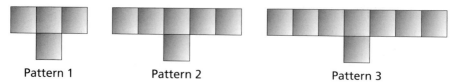

Pattern 1   Pattern 2   Pattern 3

(a) Sketch patterns 4 and 5.

(b) Explain what pattern 10 would look like.

(c) Copy and complete this table.

Pattern number	1	2	3	4	5	6
Number of tiles	4					

(d) How many tiles will be in pattern 20?

(e) Find an expression for the number of tiles in pattern n.

(f) How many tiles would be in pattern 100?

2 These tiles form square shapes.

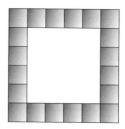

Pattern 2   Pattern 4

(a) Sketch pattern 3 and pattern 5.

(b) How many tiles are there in each of these four patterns?
Show your results in a table.

(c) Explain what pattern 20 would look like.
How many tiles would be in this pattern?

(d) Find an expression for the number of tiles in the nth pattern.

(e) One pattern has 96 tiles. Which one is it?

3   Here are some dot and line designs.
Below the designs are some formulas.

For each design

(a) Sketch the next pattern.

(b) Write down how many **dots** are added to each pattern
to make the next.

(c) Decide which of the formulas below gives the number of
dots in pattern n.

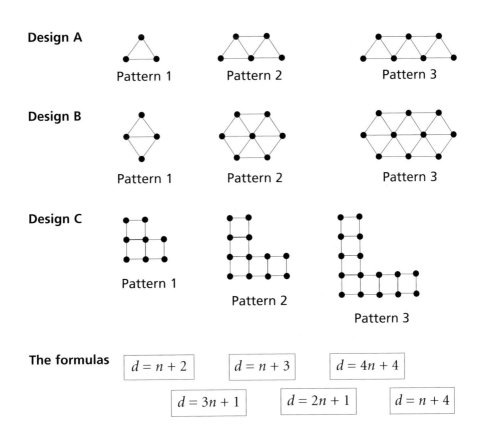

Design A   Pattern 1   Pattern 2   Pattern 3

Design B   Pattern 1   Pattern 2   Pattern 3

Design C   Pattern 1   Pattern 2   Pattern 3

The formulas $d = n + 2$ $d = n + 3$ $d = 4n + 4$

$d = 3n + 1$ $d = 2n + 1$ $d = n + 4$

Sections C and D

1

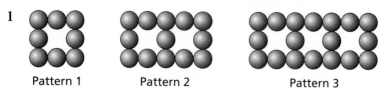

Pattern 1   Pattern 2   Pattern 3

(a) Count the number of beads in each pattern.

(b) How many beads will be in pattern 4?

(c) How many beads are added to each pattern to make the next?

(d) Find a formula for the number of beads in the nth pattern.

(e) Use your formula to work out the number of beads in pattern 30.

2

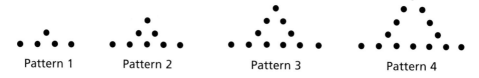

Pattern 1   Pattern 2   Pattern 3

(a) Count the number of **lines** in pattern 3.

(b) Sketch pattern 4 and count the number of lines in this pattern

(c) How many lines are added to each pattern to make the next?

(d) Find a formula for the number of lines in the nth pattern.

(e) Use your formula to work out the number of lines in pattern 10.

3 Find a formula for the number of dots in the nth pattern.

Pattern 1   Pattern 2   Pattern 3   Pattern 4

32

4 Here are tables for some more designs.
For each table, work out a formula for the number of lines in the *n*th pattern.

(a)

Pattern number (*n*)	1	2	3	4	5
Number of lines (*l*)	7	9	11	13	15

(b)

Pattern number (*n*)	1	2	3	4	5
Number of lines (*l*)	7	12	17	22	27

5 This pattern is made with square and triangular tiles.

   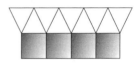

Pattern 1   Pattern 2   Pattern 3   Pattern 4

(a) Find an expression for the number of triangles in pattern *n*.

(b) Work out the number of triangles in pattern 10.

(c) One of these patterns has 241 triangles in it.

    (i) Form an equation and solve it to find *n*.

    (ii) How many squares are in this design?

6 (a) Find an expression for the number of **lines** in the *n*th pattern.

  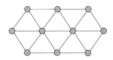

Pattern 1   Pattern 2   Pattern 3

(b) One of these patterns has 40 lines in it.
Form an equation and solve it to find *n*.

⑪ Estimation

Sections A, B, C and D

1 (a) Estimate the length and width of your living room to the nearest $\frac{1}{2}$ metre.

  (b) Roughly what is the floor area of your living room?

  (c) If fitted carpet was £24.99 a square metre, roughly what would it cost to carpet the room?

2 A cubit is the distance from your elbow to your fingertips. It is roughly 50 cm.

  (a) Estimate the length of a car in cubits.

  (b) Roughly how long is this in metres?

3 What is missing from each statement?

  (a) 8 kilometres is equivalent to about _____ miles.

  (b) 2 inches is equivalent to about _____ centimetres.

  (c) A metre is a bit longer than _____ feet.

  (d) 2 centimetres is a bit less than _____ .

4 These are the distances between some major places in Italy. Convert these into miles.

Bologna to Venice	160 km	Florence to Genoa	240 km
Rome to Naples	200 km	Florence to Bologna	100 km

5 These are the distances between some major places in the UK. Convert these into kilometres.

Lincoln to Birmingham	90 miles	York to Dundee	250 miles
Leeds to Dover	260 miles	Edinburgh to London	390 miles

6 Here are the heights of some pupils in inches. How tall are they in centimetres?

David	60 inches	Sarah	52 inches
Diana	49 inches	Tim	68 inches

7   In a shopping centre, you need to be between 3 feet
    and $4\frac{1}{4}$ feet tall to play in the children's play area.

    Jack is 1.2 metres tall.
    Can he use the play area?
    Explain your answer.

8   Write down the missing number from each statement.

    (a) A litre is a bit less than _____ pints.

    (b) A gallon is a bit more than _____ litres.

9   The fuel tank capacities of various vehicles are given here in gallons.

    How many litres does each tank hold?

    Volvo 340   10 gallons
    Cherokee Jeep   13 gallons
    Dennis Dart bus   48 gallons

10 Decide which of these statements is true.

    A   One pound is a bit more than $\frac{1}{2}$ kg.

    B   One pound is a bit less than $\frac{1}{2}$ kg.

11 Mrs Smith needs 2 pounds of sugar to make jam.
    Will a 1 kg bag be enough?

12 John weighs 75 kg and Prakash weighs 160 pounds.
    Who is heavier?

13 Choose the correct metric unit on the right for each statement.

    (a) My pet mouse weighs 13 _____ .

    (b) My big toe nail is 17 _____ wide.

    (c) My car's fuel tank holds 45 _____ .

    (d) My car is 1.5 _____ wide.

    (e) My dog weighs 6.3 _____ .

    (f) The length of my foot is 24 _____ .

    (g) I often cycle 3 _____ to the local shop.

    (h) My favourite cake recipe uses a 5 _____ spoonful of vanilla essence.

    (millimetres)   (metres)

    (grams)   (kilograms)

    (centimetres)   (kilometres)

    (millilitres)   (litres)

⑫ Quadrilaterals

Section A

1 Copy these on to square dotty paper and complete them.

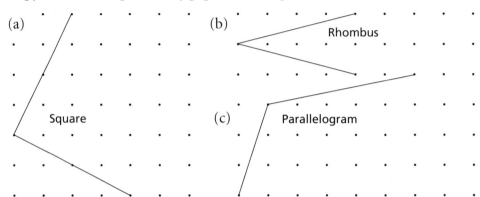

(a) Square

(b) Rhombus

(c) Parallelogram

Section B

1 For this question, work on square dotty paper if you like.
Do drawings to show how

(a) a square can be split into two trapeziums

(b) a rhombus can be split into two parallelograms

(c) a kite can be split into a rhombus and an arrowhead

(d) a rectangle can be split into two trapeziums and a parallelogram

(e) a square can be split into two arrowheads and a rhombus

Section C

1 Work out the angles marked with letters.

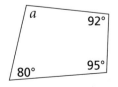

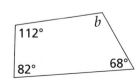

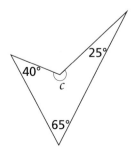

2 Work out the angles marked with letters.

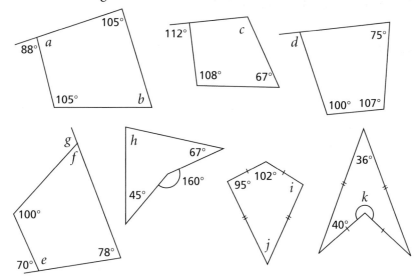

3 Work out the angles marked with letters.

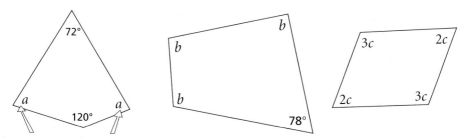

These two are the same size.

Section D

1 Do accurate drawings of the quadrilaterals sketched here.

(a)

(b)

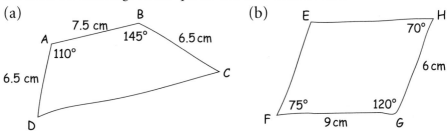

⑬ Negative numbers

Sections A and B

1 Work these out.

(a) $3 + {}^-2$ (b) $5 + {}^-8$ (c) ${}^-4 + {}^-2$ (d) ${}^-9 + 7$

(e) $6 + {}^-10$ (f) ${}^-4 + {}^-3$ (g) ${}^-2 + {}^-9$ (h) ${}^-8 + {}^-4$

2 Copy and complete these addition walls.

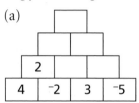 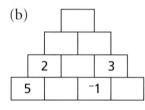 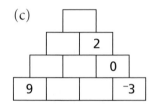

3 Work these out.

(a) $4 - 2$ (b) ${}^-5 - 1$ (c) ${}^-3 - 3$ (d) $6 - {}^-1$

(e) $8 - {}^-3$ (f) $0 - {}^-1$ (g) ${}^-4 - {}^-3$ (h) ${}^-5 - {}^-7$

4 Copy and complete these **subtraction** strips.

(a) | 10 | 8 | | | |

(b) | 5 | 7 | | | |

(c) | 8 | ⁻3 | | | |

(d) | ⁻4 | ⁻2 | | | |

(e) | 12 | | ⁻5 | | |

(f) | | ⁻3 | | ⁻1 | |

5 For each calculation, fit these numbers
into the squares to make it true.

     ${}^-4$ ${}^-2$ 1

(a) ☐ + ☐ − ☐ = ${}^-7$

(b) ☐ − ☐ − ☐ = 1

(c) ☐ + ☐ − ☐ = ${}^-1$

(d) ☐ − ☐ − ☐ = 7

Sections C and D

1 Copy and complete these multiplication squares.

(a)

×	3	5
2		
$^{-}1$		

(b)

×	4	$^{-}3$
$^{-}2$		
$^{-}5$		

(c)

×		$^{-}1$
6	$^{-}12$	
$^{-}4$		

(d)

×	$^{-}7$	
5		$^{-}25$
	28	

2 Work these out.

(a) $^{-}2 \times 3 \times {}^{-}4$ (b) $^{-}3 \times {}^{-}3 \times {}^{-}3$ (c) $2 \times {}^{-}3 \times 5$ (d) $^{-}5 \times {}^{-}2 \times 2$

3 Copy and complete these multiplication walls.

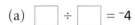

(a)

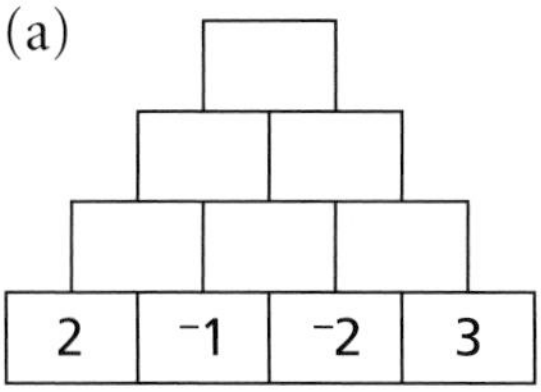

(b)

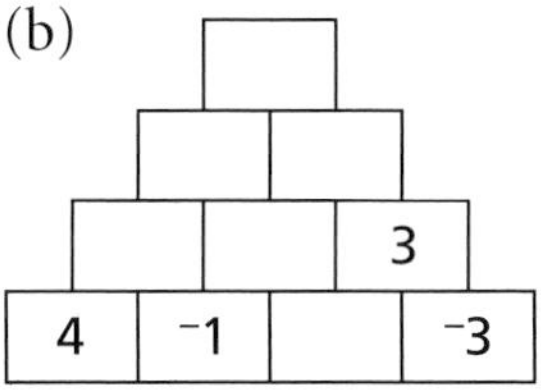

(c)

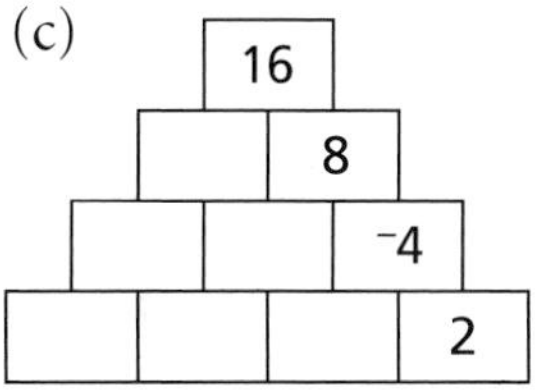

4 Fit numbers from the cloud into the squares to make each calculation true.
There are two answers for each one.

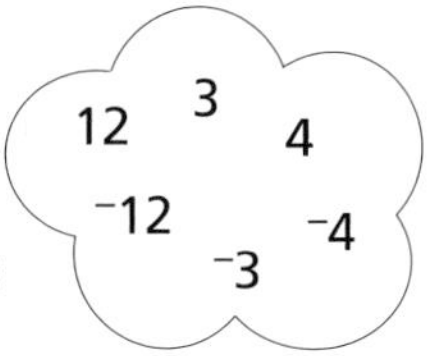

(a) ☐ ÷ ☐ = $^{-}4$ (b) ☐ ÷ ☐ = 3

(c) ☐ ÷ ☐ = 4 (d) ☐ ÷ ☐ = $^{-}3$

5 Work these out.

(a) $\frac{56}{^{-}7}$ (b) $\frac{^{-}30}{^{-}6}$ (c) $\frac{^{-}54}{9}$ (d) $\frac{^{-}49}{^{-}7}$

6 Work these out.

(a) $4 \times {}^{-}3$ (b) $^{-}8 \div 2$ (c) $^{-}7 \times {}^{-}6$ (d) $^{-}20 \div 4$

(e) $6 \times {}^{-}6$ (f) $^{-}12 \div {}^{-}3$ (g) $^{-}3 \times {}^{-}8$ (h) $32 \div {}^{-}8$

(i) $5 \times {}^{-}7$ (j) $^{-}27 \div {}^{-}9$ (k) $^{-}6 \times 9$ (l) $24 \div {}^{-}6$

Section E

1 Work these out.

(a) $2 + {}^-7$
(b) ${}^-8 + {}^-6$
(c) $4 - {}^-3$
(d) ${}^-1 - {}^-6$

(e) ${}^-3 \times {}^-5$
(f) $6 \times {}^-2$
(g) ${}^-16 \div {}^-2$
(h) ${}^-32 \div 4$

2 Work these out.

(a) $\dfrac{{}^-2 + 4}{2}$
(b) $\dfrac{{}^-8}{{}^-3 + 1}$
(c) ${}^-4({}^-2 - {}^-3)$
(d) $6(2 - {}^-8)$

(e) $\dfrac{{}^-3 - {}^-5}{{}^-2}$
(f) $\dfrac{15}{{}^-3} + {}^-2$
(g) $({}^-4)^2 + 3$
(h) ${}^-4\left(\dfrac{8}{{}^-2} - 3\right)$

3 Work these out when $x = {}^-5$.

(a) $x + 3$
(b) $2x - 1$
(c) $4 - x$
(d) $x^2 + x$

4 Work these out when $r = {}^-4$ and $s = {}^-6$.

(a) $r + s$
(b) $r - s$
(c) $s - r$
(d) $\dfrac{s}{r}$

(e) $r(s + 2)$
(f) $\dfrac{sr}{8}$
(g) $s^2 + 2r$
(h) $r(1 - s)$

5 Copy and complete these **addition** strips.

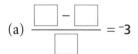

(a) | | 4 | | ${}^-1$ | |

(b) | | | 4 | | ${}^-6$ |

(c) | | | | ${}^-5$ | ${}^-6$ |

6 Fit numbers from the cloud into the squares
to make each calculation true.
There is more than one answer for each one.

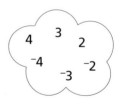

4   3   2 ${}^-4$ -2   -3

(a) $\dfrac{\square - \square}{\square} = {}^-3$

(b) $\square\,(\square + \square) = 4$

(c) $\dfrac{\square}{\square - \square} = {}^-4$

⑭ Percentage change

Section A

Do not use a calculator for these questions.

1 What is the output for each flow diagram?

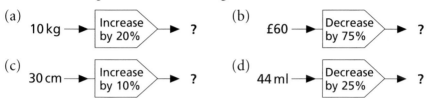

(a) 10 kg → Increase by 20% → ?

(b) £60 → Decrease by 75% → ?

(c) 30 cm → Increase by 10% → ?

(d) 44 ml → Decrease by 25% → ?

2 What is the missing percentage in each flow diagram?

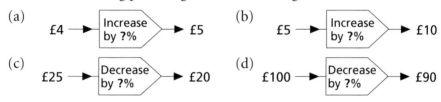

(a) £4 → Increase by ?% → £5

(b) £5 → Increase by ?% → £10

(c) £25 → Decrease by ?% → £20

(d) £100 → Decrease by ?% → £90

3 What is the final result for each flow diagram?

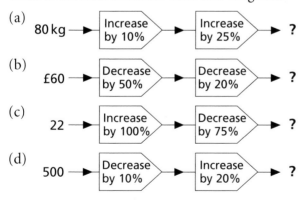

(a) 80 kg → Increase by 10% → Increase by 25% → ?

(b) £60 → Decrease by 50% → Decrease by 20% → ?

(c) 22 → Increase by 100% → Decrease by 75% → ?

(d) 500 → Decrease by 10% → Increase by 20% → ?

*4 What is the missing amount in each flow diagram?

(a) ? → Increase by 20% → £60

(b) ? → Decrease by 25% → £60

Section B

1 (a) Which of these multipliers will increase an amount by 37%?

    × 0.37   × 1.37   × 137   × 100.37   × 37

    (b) Increase £70 by 37%.

2 What is the output for each flow diagram?

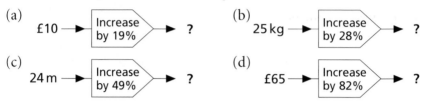

(a) £10 → Increase by 19% → ?

(b) 25 kg → Increase by 28% → ?

(c) 24 m → Increase by 49% → ?

(d) £65 → Increase by 82% → ?

3 John starts an exercise programme to increase his weight by 18%.
Before he start the programme his weight is 75 kg.

What weight is he aiming for?

4 A restaurant increases its prices by 12%.

    (a) A pizza previously cost £5.00.
        What is the cost of a pizza after the increase?

    (b) A steak previously cost £10.50.
        What is the new cost of a steak?

5 (a) Increase £20 by 36%.   (b) Increase 65 kg by 74%.

    (c) Increase 5.6 m by 15%.   (d) Increase £256 by 27%.

    (e) Increase 48 kg by 45%.   (f) Increase 25 m by 6%.

Section C

1 (a) Which of these multipliers will decrease an amount by 37%?

    × 0.37   × 1.63   × 1.37   × 63   × 0.63

    (b) Decrease £70 by 37%.

2 What is the output for each flow diagram?

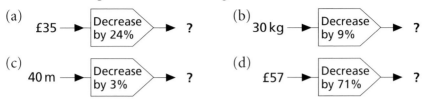

(a) £35 → Decrease by 24% → ?

(b) 30 kg → Decrease by 9% → ?

(c) 40 m → Decrease by 3% → ?

(d) £57 → Decrease by 71% → ?

3 In a sale a shop reduces all its prices by 33%.
Reduce each of these prices by 33% and give
the new prices to the nearest penny.

(a) £25   (b) £83   (c) £18

(d) £6.70   (e) £39.90   (f) £84.60

Section D

1 Every percentage increase or decrease corresponds to a multiplier.
Match these percentage changes to their multipliers.

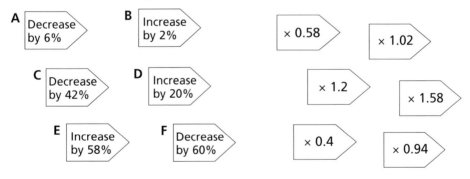

A Decrease by 6%   B Increase by 2%   × 0.58   × 1.02

C Decrease by 42%   D Increase by 20%   × 1.2   × 1.58

E Increase by 58%   F Decrease by 60%   × 0.4   × 0.94

2 (a) Increase £70 by 46%.   (b) Decrease £150 by 14%.

    (c) Decrease 60 kg by 8%.   (d) Increase 40 kg by 35%.

    (e) Decrease 3400 by 74%.   (f) Increase 9000 by 9%.

3 Jane buys a car priced at £8500.
She is given a discount of 5% for paying cash.

How much does she pay for her car?

Mixed questions 2

1 Which is slower, a speed of 30 mph or a speed of 50 km per hour? Explain your answer.

2 The number of houses in a village is due to increase by 8% over the next year. If there are already 2450 houses, how many will there be in a year's time?

3 (a) Estimate the length of the longest side of this book in centimetres.

   (b) About how many of these books, laid end to end, would cover 2 metres?

4 Here are some patterns of crosses.

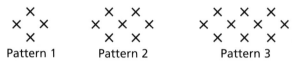

Pattern 1   Pattern 2   Pattern 3

   (a) How many crosses would be needed for pattern 10?

   (b) Find a formula for the number of crosses in the nth pattern.

   (c) One pattern needs 100 crosses. Which one is it?

5 Calculate the areas of these flower-beds.

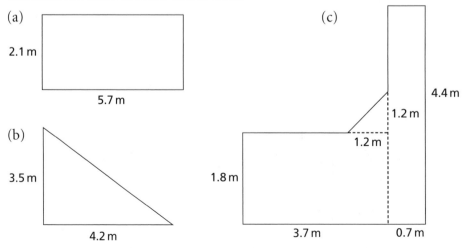

6 (a) Ron has a 5 gallon fuel can. About how many litres will this hold?

(b) Marty's bike has wheels with a 26 inch diameter. About how many centimetres is this?

7 Work these out and put them in order, lowest first, to find a word.

$^-5 \times ^-2$	$(^-2)^2$	$\dfrac{4-6}{2}$	$3(6-^-2)$	$^-6-^-9$	$\dfrac{^-2 \times ^-8}{^-4}$	$\dfrac{^-12}{^-2+4}$	$8-11$
A	**N**	**G**	**L**	**O**	**I**	**D**	**A**

8 Here is a table for the number of counters needed for a design.

Pattern number (n)	1	2	3	4	5
Number of counters (c)	2	8	14	20	26

(a) Find a formula for the number of counters in the nth pattern.

(b) Use your formula to find how many counters would be needed for the 20th pattern.

(c) Which pattern needs 86 counters?

9 Draw these quadrilaterals accurately.

(a)

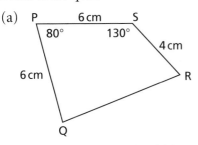

(b)

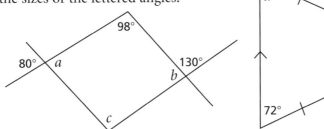

(c) Measure the length of (i) QR (ii) VW

10 Find the sizes of the lettered angles.

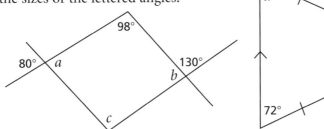

11 Find the missing lengths.

(a)

Area = 4.2 m²   1.5 m

a m

(b)

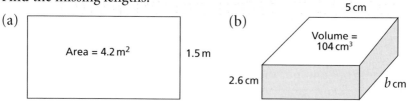

5 cm

Volume = 104 cm³

2.6 cm

b cm

12 Use these three numbers to make these calculations correct.

(a) $\square - \square + \square = 7$

(b) $\square - \square + \square = 3$

$^-4 \quad ^-2 \quad 5$

(c) What is the lowest number you can make using the three numbers
and $\square - \square + \square = $?

13 The price of £75 for a CD player is due to increase by 6% next week.
How much would you have to pay for the CD player next week?

14 Find the quadrilateral.

Clue A:	One diagonal divides the quadrilateral into two congruent obtuse-angled scalene triangles.

(a) What types of quadrilateral could it be?
You should be able to name three possibilities.

Clue B:	The other diagonal divides the quadrilateral into two isosceles triangles.

(b) Which type of quadrilateral must it be?

15 The value of a computer decreased by 22% over the last year.
It was worth £1500 a year ago.

How much is it worth now?

⑮ Probability from experiments

Section B

1 Maureen works in a dress factory.
   She has to reject any dress that is not perfect.
   One day she checks 200 dresses and rejects 30 of them.
   Estimate the probability that the next dress she checks is perfect.

2 Gary stands at a crossroads and does a survey of traffic that has
   come from town. He records whether vehicles turn left, turn right
   or go straight on. These are his results.

   R R S R L R S S L R L L S R L
   L S L L R S S R R L R S S L R

   What is the relative frequency of a vehicle
   (a) turning left   (b) turning right   (c) going straight on

3 Zara and Naomi enjoy playing chess together.
   They record who wins each game. Here are the results.

   Z N N Z Z Z Z N Z N N Z Z N N N Z N Z Z
   N N Z N Z Z N N N N N N Z Z N Z Z Z N

   What is the relative frequency of
   (a) Naomi winning   (b) Zara winning

   Who seems to be the better chess player?

4 Jasmine runs the school tuck shop.
   She records the flavours of crisps
   that are sold one break time.

   Here is her record at the
   end of break.

Flavour	Tally
Ready salted	ＨＨＴ ＨＨＴ ＩＩ
Salt and vinegar	ＨＨＴ ＨＨＴ ＨＨＴ ＩＩＩ
Cheese and onion	ＨＨＴ ＨＨＴ ＨＨＴ ＨＨＴ Ｉ
Prawn cocktail	ＨＨＴ ＨＨＴ ＨＨＴ ＨＨＴ ＩＩＩ
Beef	ＨＨＴ ＩＩ
Pickled onion	ＨＨＴ ＨＨＴ

   Work out the relative frequency for each flavour.
   Write them as decimals, to two decimal places.

Section C

1   A fair spinner is marked like this.

If the arrow is spun 100 times, roughly how many times would you expect it to land on the section marked

(a) crown   (b) anchor   (c) star

2   An ordinary dice is thrown 500 times.

Roughly how many times would you expect it to show

(a) five   (b) an odd number   (c) a multiple of 3

(d) a square number   (e) a factor of 12

3   Chris travels to work on the bus for 70 days.
    He arrives late on 23 of the days.

(a) Estimate the probability that Chris is late for work.

(b) If Chris travels to work on the bus for 200 days,
    roughly how many times should he expect to be late?

4   Over the past few days, a shop sold the
    following numbers of packets of crisps.

Ready salted	83
Salt and vinegar	52
Cheese and onion	41
Prawn cocktail	27
Beef	19
Pickled onion	13

(a) Estimate the probability that the next packet sold is
    ready salted.

(b) The shop is going to order 1000 packets of crisps.
    About how many of each flavour should they order?

⑯ Squares, cubes and roots

Sections A and B

Do not use a calculator for these questions.

1 Write down the value of
   (a) the square of 7   (b) 4 squared
   (c) 3 cubed   (d) the cube of 1

2 Find a square number and a cube number which add to give 72.

3 Janique has 52 square slabs.
   She arranges them to make two square patios.
   What are the sizes of the patios, if she has used all of the slabs?

4 Write down two square numbers which are also cube numbers.

5 Work these out.
   (a) $(^-4)^2$ (b) $(^-4)^3$ (c) $(^-5)^2$ (d) $(^-5)^3$

6 (a) Use the fact that $45^2 = 2025$ to write down the value of $(^-45)^2$.
   (b) (i) Write down the positive square root of 2025.
        (ii) Write down the negative square root of 2025.

7 What is the positive square root of 81?

8 What is the negative square root of 144?

9 Find two numbers that fit each statement
   (a) $\blacksquare^2 = 64$ (b) $\blacksquare^2 = 36$ (c) $\blacksquare^2 = 100$ (d) $\blacksquare^2 = 49$

10 Copy and complete this.

> 4 cubed is $4^3 = 4 \times 4 \times 4 = \blacksquare$
> so the cube root of $\blacksquare$ is 4.
>
> $(^-4)$ cubed is $(^-4)^3 = \blacksquare \times \blacksquare \times \blacksquare = \blacksquare$
> so the cube root of $\blacksquare$ is $^-4$.

11 Find the cube root of

    (a) $^-1$ (b) 125   (c) $^-8$ (d) 1000

Use a calculator for these questions.

12 Find two square numbers between 700 and 800.

13 Work these out.

    (a) 2.7^2 (b) 4.1^3 (c) 0.6^2 (d) 1.8^3

14 Find a number that fits each statement

    (a) $\blacksquare^2 = 7.84$ (b) $\blacksquare^3 = 4.913$ (c) $\blacksquare^2 = 18.49$ (d) $\blacksquare^3 = 0.512$

15 Find the cube root of

    (a) 729   (b) 2744   (c) 0.216   (d) $^-343$

Section C

1 (a) What is the value of 5.4^2?

    (b) What is the value of 5.5^2?

    (c) Why do your results show that the positive square root of 30 is between 5.4 and 5.5?

    (d) Find the positive square root of 30 correct to 2 d.p.

    (e) What is the negative square root of 30 correct to 2 d.p.?

2 (a) What is the value of 5.8^3?

    (b) What is the value of 5.9^3?

    (c) Why do your results show that the cube root of 200 is between 5.8 and 5.9?

    (d) Find the cube root of 200 correct to 2 d.p.

3 Find, correct to 2 d.p.

    (a) the positive square root of 40

    (b) the cube root of 20

17 Area

Section A

1 Find the area of each of these parallelograms.

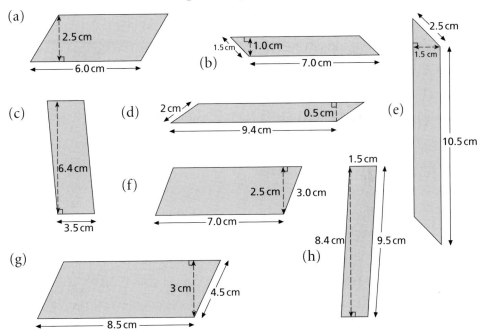

(a) 2.5 cm, 6.0 cm

(b) 1.5 cm, 1.0 cm, 7.0 cm

(e) 2.5 cm, 1.5 cm, 10.5 cm

(c) 6.4 cm, 3.5 cm

(d) 2 cm, 0.5 cm, 9.4 cm

(f) 2.5 cm, 3.0 cm, 7.0 cm

(h) 1.5 cm, 8.4 cm, 9.5 cm

(g) 3 cm, 4.5 cm, 8.5 cm

2 The area of each of these parallelograms is 36 cm².
Find the missing length in each one.

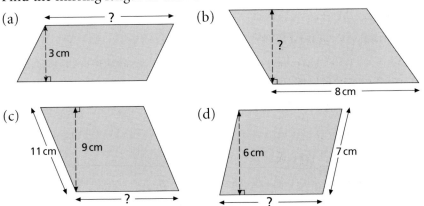

(a) ?, 3 cm

(b) ?, 8 cm

(c) 11 cm, 9 cm, ?

(d) 6 cm, 7 cm, ?

51

Section B

1 Put these triangles into groups so that the triangles in each group have the same area.

For each group say

- which triangles are in it
- what the areas of the triangles are

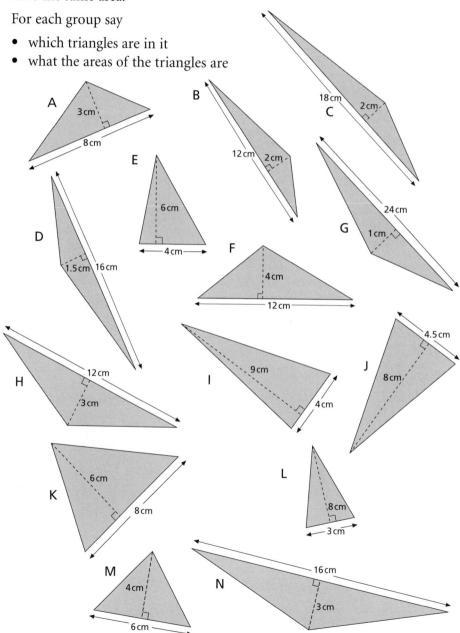

2   Find the area of each of these triangles.
    They are not drawn accurately.

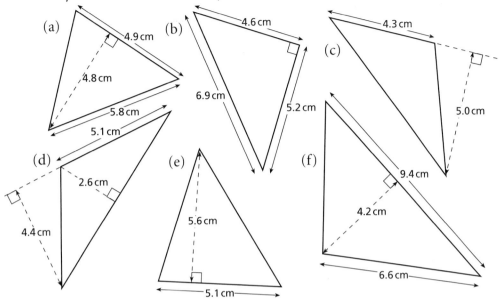

(a) 4.9 cm 4.8 cm 5.8 cm 5.1 cm

(b) 4.6 cm 6.9 cm 5.2 cm

(c) 4.3 cm 5.0 cm

(d) 2.6 cm 4.4 cm

(e) 5.6 cm 5.1 cm

(f) 9.4 cm 4.2 cm 6.6 cm

Section C

1   Put these trapeziums into groups so that the
    trapeziums in each group have the same area.

    For each group say

    • which trapeziums are in it

    • what the area of each trapezium is

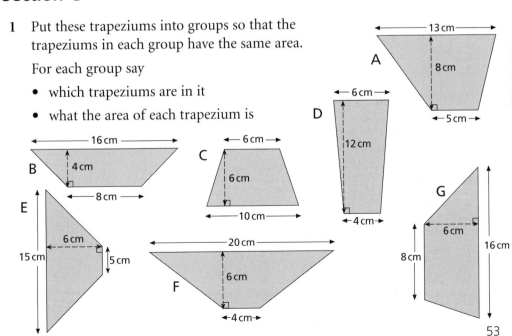

A   13 cm   8 cm   5 cm

D   6 cm   12 cm   4 cm

B   16 cm   4 cm   8 cm

C   6 cm   6 cm   10 cm

E   6 cm   15 cm   5 cm

F   20 cm   6 cm   4 cm

G   6 cm   8 cm   16 cm

53

Equivalent expressions

Section B

1 (a) Copy and complete this wall.

(b) What number will be on the top brick when $n = 2$?

(c) What value for n will give 20 on the top brick?

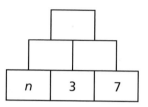

2 (a) Copy and complete these walls.

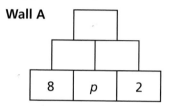

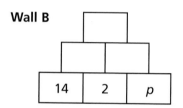

(b) For each wall work out the number on the top brick

(i) when $p = 6$

(ii) when $p = 7$

3 Copy and complete these walls.

(a)

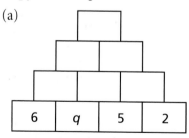

(b)

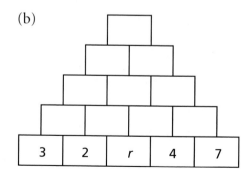

4 Copy and complete these walls.

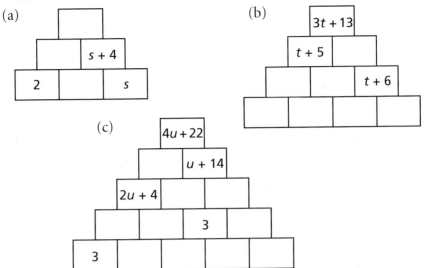

(a)

| | |
| 2 | | s |

with s + 4

(b)

3t + 13
t + 5
t + 6

(c)

4u + 22
u + 14
2u + 4
3
3

5 Simplify these expressions.

(a) $a + 2 + a + 4$ (b) $b + 3 + 2b$ (c) $x + x + 5 + x$

(d) $3y + 4 + 2y + 6$ (e) $2p + 6 + 3p + 2p + 4$ (f) $j + 3j + 5$

(g) $5g + 8 + g + 2g$ (h) $q + 2q + 6 + q + 5$

*6 Which pair of these walls will share the same top brick number, whatever the value of p?

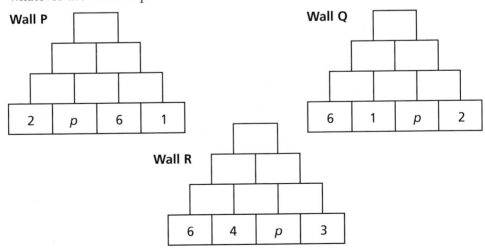

Wall P

| 2 | p | 6 | 1 |

Wall Q

| 6 | 1 | p | 2 |

Wall R

| 6 | 4 | p | 3 |

Section C

1. Write an expression, in its simplest form, for the perimeter of each of these shapes.

(a)
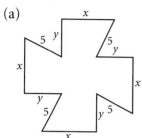

(b)
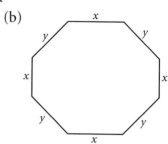

(c)
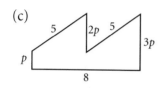

(d)
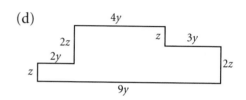

2. Simplify the following.

(a) $3g + g + 2h + h$

(b) $8p + 2q + 3q + 2p$

(c) $4m + 5 + 2m + n$

(d) $4 + 2t + 8 + u + 4t$

(e) $5 + a + 3a + 4b + 5a$

(f) $3 + x + y + 2x + 4 + 5y$

3. Sketch each shape and write expressions for the missing lengths.

(a)
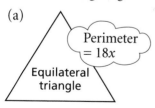

Perimeter = $18x$

Equilateral triangle

(b)
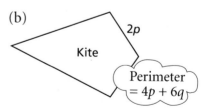

$2p$

Kite

Perimeter = $4p + 6q$

Section D

1   Simplify the following.

(a) $4p + 6 - 2p + 3$ (b) $3q + 6 - 2q - 2$

(c) $8r + 4 + r - 6 + 2r$ (d) $5s - 3 + 2s + 6$

(e) $3t - 5 - 2t + 9$ (f) $3u - 4 - 2u + 6 + 3u$

2   Find the matching pairs.

A $4p + 2 - p$ **B** $3p + 6 - p - 5$ **C** $2p + 6 + p - 2$

D $2p + 5 + p - 1$ **E** $2 + 6p - 4p - 1$ **F** $2p + 5 - 3 + p$

3   Simplify the following:

(a) $3v + 2w - v + 6w$ (b) $5 + 4p - 3 - 2p$

(c) $2 + 6a - 4a + 3$ (d) $2q + 5r + 3q - 6r$

(e) $8t + 6u - 4t - 8u$ (f) $3 - 4y + 2y + 7$

4   Which of these can be simplified to $a - b$?

A $3a + 2b - 2a + b$ **B** $2a - 5b - a + 6b$ **C** $3a - 2b + b - 2a$

5   Simplify the following.

(a) $4p + 2q - 7q + 2p$ (b) $8 - 2r + s - 3r + 2s$

(c) $14 - 6t + 3u - 4 + 5t$ (d) $4v - 3w - 2v - 8w$

(e) $5a - 6 + 3a - 5$ (f) $6p - q - 5p + 2q$

(g) $3x + 2y - 2x - 2y$ (h) $3g - 2h + 4g - h + 3h$

*6   Write an expression for the missing length.
Simplify your expression.

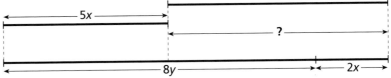

57

Section E

1 (a) This grid always gives a magic square
Explain why.

(b) Make a magic square using $p = 4$.

(c) What value of p gives a
total of 30 on each line?

$4p + 1$	$2 + 9p$	$2p - 3$
$3p - 4$	$5p$	$4 + 7p$
$8p + 3$	$p - 2$	$6p - 1$

2 Which of these always gives a magic square?

Square A

$3a + b$	$2a - b$	$4a$
$4a - b$	$3a$	$b + 2a$
$2a$	$4a + b$	$3a - b$

Square B

$10x + y$	$9x - 4y$	$5x$
$8x - y$	$4x + 3y$	$12x - 5y$
$6x - 3y$	$11x - 2y$	$7x + 2y$

3 (a) Write in its simplest form an
expression for the total of each
line on this magic square.

$17 - k$	$12 - 6k$	$4k + 19$
$4k + 18$	$16 - k$	$14 - 6k$
$13 - 6k$	$20 + 4k$	$15 - k$

(b) Use this grid to make a magic square with

(i) 14 as the middle number

(ii) a total of 45 on each line

Section F

1 Multiply out the brackets from these.

(a) $4(p + 5)$ (b) $3(q + 2)$ (c) $6(3 + r)$

(d) $8(s - 2)$ (e) $5(3 - t)$ (f) $12(u - 4)$

2 Multiply out these brackets.

(a) $2(2x + 4)$ (b) $3(4 + 3y)$ (c) $5(6 + 3z)$

(d) $9(3a - 4)$ (e) $5(2b - 3)$ (f) $11(3 - 2c)$

3 Copy and complete these statements.

(a) $4(d + 3) = 4d + \square$ (b) $3(5 - 2e) = 15 - \square$

(c) $8(\square - 2) = 8f - 16$ (d) $4(\square - 9) = 4g - 36$

4 Which two expressions give the perimeter of this shape?

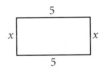

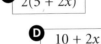

Ⓐ $10 + x$ **Ⓑ** $2(5 + 2x)$

Ⓒ $2(5 + x)$ **Ⓓ** $10 + 2x$

5 Find three pairs of equivalent expressions.

Ⓐ $3(x + 3y)$ **Ⓑ** $3x + 3y$ **Ⓒ** $3(x + y)$

Ⓓ $3x + y$ **Ⓔ** $3x + 9y$ **Ⓕ** $y + 3x$

6 Find two expressions for the perimeter of this shape.

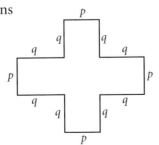

⑲ No chance!

Section A

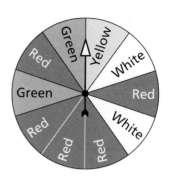

If this Wheel of Fortune is spun,
all sections are just as likely to win.

1 Which colour on the Wheel of Fortune
is most likely to come up?

2 (a) What is the probability of getting a yellow section?

(b) What is the probability of not getting a yellow section?

3 What is the probability of each of these 'events' happening when the wheel
is spun?

(a) getting a white section   (b) getting a red section

(c) not getting a red section   (d) getting a black section

(e) not getting a green section

4 A drawer contains 7 pink, 5 white and 8 blue socks.
If a sock is chosen randomly from the drawer,
what is the probability of it being

(a) pink   (b) not pink   (c) white

(d) green   (e) white or pink   (f) white or blue

5 A drawer contains 24 socks which are either white, black or grey.
The probability of randomly choosing a black sock is $\frac{1}{2}$,
and the probability of choosing a white sock is $\frac{1}{3}$.

(a) How many black socks are there?

(b) How many white socks are there?

(c) What is the probability of not choosing white?

(d) What is the probability of choosing a grey sock?

(e) What is the probability of not choosing a grey sock?

Section B

1   This spinner is spun twice.

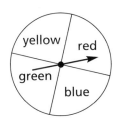

(a) List all the different pairs of colours
     that could be obtained.
     For example,

1st spin	2nd spin
R	R

(b) What is the probability that

   (i)   both spins give the same colour

   (ii) the two spins give different colours

   (iii) neither spin gives yellow

2   These are four digit cards.

1   **4**   **9**   **6**

Gabbi chooses two cards randomly and
places them to make a two-digit number,
                    for example 94

9 **4**

(a) List all the possible two-digit numbers she could make.
     (You only use each card once!)

(b) How many different numbers could she make?

(c) What is the probability that the two-digit number
     that Gabbi makes is

   (i)   an even number   (ii) a multiple of 7

   (iii) a square number   (iv) a prime number

   (v)   a factor of 96   (vi) not a prime number

Sections C and D

1   For a game two spinners are used, one a square
    and the other a regular pentagon.

    In the game the spinners are spun
    and the scores are added together.

Copy and complete the grid below to show all the possible totals.

Pentagon spinner

	1	**2**	**3**	**4**	**5**
1	2	3	...		
2	...	...	...		
3					
4					

Square spinner

2   Use your grid to find the probabilities of getting

    (a) a total of 7

    (b) a total of 2

    (c) a total of 8 or more

    (d) a double (the same number on both)

3   Which totals are you most likely to get with two spinners like this?

4   In a different game two spinners like these are used.

    In this game the scores are **multiplied** together.

    (a) Draw a grid to show all the possible scores.

    (b) Use your grid to find out which score is most likely.

5   Which of these is more likely? Explain why.

    A   Rolling a multiple of 3 on an ordinary dice

    B   Getting an odd number when picking at random from cards
            with 2, 3, 4, 5, 6, 7, 8 on them

⑳ Recipes

Sections A and B

The recipes show some of the ingredients required for different dishes.
Calculate the amount of …

1 … butter in a steak and kidney pie for 8 people.

2 … steak in a pie for 9 people.

3 … beef stock in a pie for 10 people.

4 … mushrooms in a pie for 7 people.

5 … flour in a pie for 15 people.

Steak and kidney Pie	
Serves 4	
200 g	plain flour
700 g	braising steak
175 g	ox kidney
100 g	butter
100 g	mushrooms
150 ml	beef stock
15 ml	tomato puree
15 ml	Worcestershire sauce

6 … cream in a pasta and mushroom bake for 9 people.

7 … mushrooms in a pasta and mushroom bake for 8 people.

8 … noodles in a pasta and mushroom bake for 10 people.

Pasta and mushroom bake	
Serves 3	
225 g	ribbon noodles
25 g	butter
50 g	Stilton cheese
225 g	mushrooms
60 ml	cream
100 g	Mozzarella cheese

9 Rewrite the pancake ingredients for 8 large pancakes.

10 Rewrite the pancake ingredients for 30 small pancakes

Pancakes	
5 large or 12 small pancakes	
110 g	plain flour
200 ml	milk
75 ml	water
30 ml	melted butter

11 An 8 kg bag of potatoes contains 96 potatoes.
Assume all the potatoes are the same size.

(a) How many potatoes would there be in a 1 kg bag?

(b) How many potatoes would be in a 25 kg sack?

12 A recipe for coffee fudge requires 550 g of sugar.
This makes 40 cubes of fudge.

(a) How much sugar is required for 1 cube of fudge?

(b) How much sugar would you require if you were going to make
300 cubes of fudge?

13 A small tub of Sweeties contains 31 sweets
weighing 48 grams altogether.

A large tub contains 85 sweets.

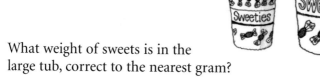

What weight of sweets is in the
large tub, correct to the nearest gram?

14 Sophie wanted to to find the weight of 1 000 000 grains of rice.
She used the science department scales and found that
30 grains of rice weighed 0.58 g.

What is the weight of 1 000 000 grains
of rice, correct to the nearest kg?

*15 Class 8R raised money for charity by filling a jar with dried peas and asking
people to guess how many peas were in the jar.

To check the answer they weighed the jar when it was empty, weighed it
when it was full of peas, and weighed 80 peas.

The empty jar weighed 550 g.
The full jar weighed 1 kg 320 g.
80 peas weighed 19 g.

How many peas were there in the full jar?
Give your answer to the nearest pea.

㉑ Substituting into formulas

Section B

1 Work out the value of each expression when $m = 4$.

(a) $3m$ (b) $4(m + 1)$ (c) $20 + m$ (d) $2(m - 3)$

(e) $12 - 2m$ (f) $20 - 3m$ (g) $5(4 - m)$ (h) $6 - \frac{m}{2}$

2 Work out the value of y in each of these rules when $x = 3$.

(a) $y = x + 7$ (b) $y = 3(x + 5)$ (c) $y = \frac{4x}{6}$

(d) $y = \frac{x + 12}{3}$ (e) $y = \frac{x}{3} + 12$ (f) $y = 2(9 - x)$

3 A new candle is 30 cm long. The formula for the length of the candle is

$$l = 30 - 4t$$

t is the time in hours that the candle has burned,
l is the height of the candle in centimetres.

(a) What is l when $t = 1$?

(b) What is l when $t = 5$?

(c) How tall is the candle after burning for 3 hours?

(d) How long will it take for the candle to burn away completely?

4 Debbie cuts the grass for neighbours.
The formula for the time she takes for each lawn is

$$t = 5 + \frac{a}{2}$$

a is the area of the lawn in square metres,
t is the time in minutes.

(a) Work out t when $a = 20$.

(b) How long will Debbie take to cut
the grass when the lawn has an area of 10 square metres?

(c) How long will she take to cut the grass when the lawn's area
is 30 square metres?

(d) Debbie has 17 minutes left in her schedule one afternoon.
What is the largest area of grass she can cut in this time?

Section C

1 Work out the value of each of these expressions when $n = 3$.

(a) n^2

(b) $6 + n^2$

(c) $3n^2$

(d) $2n^2 - 10$

(e) $(n + 1)^2$

(f) $4n^2 - 30$

(g) $\dfrac{n^2}{3}$

(h) $6 - \dfrac{n^2}{9}$

(i) $10 - n^2$

(j) $(2n - 5)^2$

(k) $(10 - n)^2$

(l) $(6 - 2n)^2$

2 Work out the value of a in each of these rules when $b = 8$.

(a) $a = 2b^2$

(b) $a = \dfrac{b^2}{4}$

(c) $a = 7 - \dfrac{b}{2}$

(d) $a = 2(9 - b)$

(e) $a = \dfrac{b^2 + 1}{5}$

(f) $a = 40 - \dfrac{b^2}{2}$

3 Here are some algebra cards.

$4p - 10$ $20 - p^2$ $\dfrac{p}{4}$ $4(p + 3)$

(a) If $p = 4$, which card will give the highest value?
     What is the value of the expression?

(b) If $p = 1$, which card will give the highest value?
     What is the value of the expression?

(c) If $p = 0$, which card will give the lowest value?
     What is the value of the expression?

4 Which of these rules give a negative value of y when $x = 5$?

$y = x^2 - 20$ $y = 3 - 2x$ $y = 60 - 2x^2$

$y = 3(x - 2)$ $y = 3(2 - x)$

Sections D and E

1   The Rugged Walk outdoor centre organises walking trips.

They always take 2 cereal bars for each person
and an extra 3 for emergencies.

(a) How many cereal bars would they take for 5 people?

(b) Copy and complete this table.

It shows p (the number of people) and c (the number of cereal bars).

p (number of people)	c (number of cereal bars)
1	
2	7
3	
4	
5	
6	

(c) Which of these is a formula for the number of cereal bars?

$c = p + 3$ $c = 2p + 3$ $c = p + 2$ $c = 3p + 2$

2   Dee's Dinners organise meals for small groups of people.
They always take 3 chocolates for each person plus 6 extra.

(a) How many chocolates would they take for 4 people?

(b) Copy and complete this table.
It shows c (the number of chocolates they take) and p (the number of people).

p (number of people)	c (number of chocolates)
2	
3	

(They don't organise meals for more than 8 people, so stop at $p = 8$.)

(c) Write down a formula connecting c (the number of chocolates they take) and p (the number of people).

Write it in the form $c = ...$

3 This square has edge
lengths of n cm.

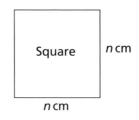

Square | n cm

n cm

(a) What is the area of the square when $n = 3$?

(b) Which of these formulas is for the area
of the square in square centimetres?

| $A = n$ | $A = n^2$ | $A = 4n$ | $A = 2n$ |

4 Iain, William and Alex are counting
the number of coins in their money boxes.

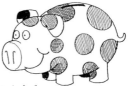

(a) Let n stand for the number of coins in Iain's box.
William has 3 more coins than Iain.
Write an expression for the number of coins William has.

(b) Alex has three times as many coins as Iain.
Write an expression for the number of Alex's coins.

(c) Sarah counts the coins in her box.
She has 5 fewer coins than Iain.
Write an expression for the number of coins Sarah has.

5 (a) Henry starts the year with 100 coins in his box.
He spends some of them. Let n stand for the number of coins spent.
Write an expression for the number of coins in Henry's box.

(b) Shelagh also starts the year with 100 coins.
She spends twice as many as Henry.
Write an expression for the number of coins left in Shelagh's box.

Mixed questions 3

1 Which two square numbers add together to make 100?

2 Simplify these expressions.

    (a) $2a + 3b - a - b$ (b) $5x - 2y + 4 - x - y$ (c) $4t - 9 - 3t - 8$

3 (a) Find and simplify an expression for the perimeter of this shape.

    (b) What value of r gives a perimeter of 60?

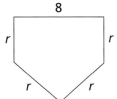

4 In these strips the value of each box is found by adding together the values in the two previous boxes.

    (a) Copy and complete this strip.
        You should find that the last box has value 28.

1	5	6			

    (b) (i) Copy and complete this strip.

n	1				

        (ii) If $n = 2$ what is the value of the last box?

        (iii) If the value of the last box is 17, what is the value of n?

5 (a) Amanda has n CDs. Lisa has three times as many CDs as Amanda.
        Write down an expression for the number of CDs Lisa has.

    (b) Greg has two fewer CDs than Lisa.
        Write down an expression for the number of CDs Greg has.

    (c) Adam has twice as many CDs as Greg.
        Write down an expression for the number of CDs Adam has.

6 Write these expressions without brackets.

    (a) $4(x + 3)$ (b) $5(2r - 3t)$ (c) $6(3a - 7)$

7 Find the areas of these triangles.

(a)

(b)

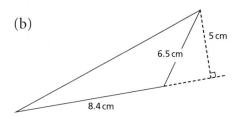

8 (a) A packet of seeds claims to have a 95% germination rate.
How many seeds would you expect to germinate in a packet of 120 seeds?

(b) For another packet of seeds the probability of germination is $\frac{8}{9}$.
How many of these seeds would you expect to germinate in a packet
of 120 seeds?

9 Write these numbers in order of size, smallest first.

$(^-2)^2$ 5^3 1^2 $(^-7)^2$ $(^-10)^3$ 10^2 $(^-2)^3$

10 (a) What is the size of angle A in this parallelogram?

(b) Draw the parallelogram accurately.

(c) Work out its area as accurately as you can.

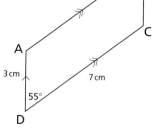

11 Find the value of each of these expressions if $r = 5$ and $t = {}^-2$.

(a) $3(r + 1)$ (b) $r + t$ (c) $2r^2$ (d) $2(1 - r)$

(e) t^2 (f) $9 - t$ (g) $\frac{r + 7}{2}$ (h) $(2r - 1)^2$

12 When Hannah makes pancakes she cooks three for each person and
an extra four. Write down a formula for the number of pancakes
she would make for n people.

13 Here is part of a recipe for spinach and lentil roulade.

<table>
<tr><td colspan="2">Spinach and Lentil Roulade</td></tr>
<tr><td colspan="2">serves 4</td></tr>
<tr><td>120 g</td><td>frozen spinach</td></tr>
<tr><td>75 g</td><td>butter</td></tr>
<tr><td>50 g</td><td>flour</td></tr>
<tr><td>300 ml</td><td>milk</td></tr>
<tr><td>175 g</td><td>lentils</td></tr>
<tr><td>2</td><td>eggs</td></tr>
</table>

(a) How much frozen spinach would be needed for 6 people?

(b) What weight of lentils would be needed for 6 people?

(c) How much milk would be needed for 10 people?

(d) Write the ratio of the weight of butter to lentils in its simplest form.

14 Roy experiments with dropping matchsticks on to a 2 cm grid.
He records the number of lines of the grid each matchstick crosses.

Here are his results.

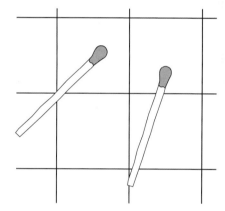

Number of lines crossed	Tally	Frequency
0		0
1	‖	2
2	卌 卌 ‖‖‖‖	14
3	卌 卌 卌 卌 ‖	21
4	‖‖‖	3
	Total	40

(a) Work out the relative frequency for each number of lines crossed. Write them as decimals. Check that they add up to 1.

(b) If the matchstick were dropped 250 times, approximately how many times would it cross two lines?

15 Find the probability of each of these events and say which is more likely.

A   Exactly two of the coins show heads when three coins are tossed.

B   Getting a total score of 8 or more with two 6-sided dice.

㉒ Scaling

Sections A, B and C

1 Copy this shape on to squared paper.

Now draw an enlargement of it using scale factor 2.

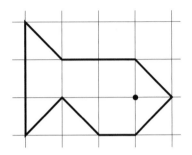

2 (a) What can you say about the angles in these two shapes?

    (b) Is the larger shape an enlargement of the smaller one?

3 (a) What can you say about the angles in these two shapes?

    (b) Is the larger shape an enlargement of the smaller one?

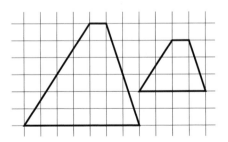

4 What is the scale factor of each of these?

    (a) A scaling from P to Q

    (b) A scaling from Q to P

P

Q

5 What is the scale factor of each of these?

   (a) A scaling from R to S

   (b) A scaling from S to R

 R S

6 These are sketches of some right-angled triangles.
They are not drawn accurately.

Which triangles are enlargements of which?

Give the scale factors.

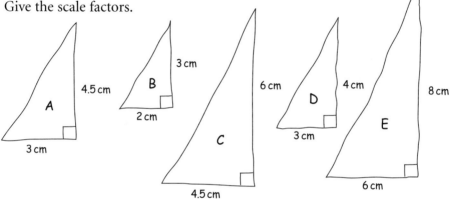

7 These are sketches of a symmetrical logo to be drawn to different scales.
Work out the missing measurements.

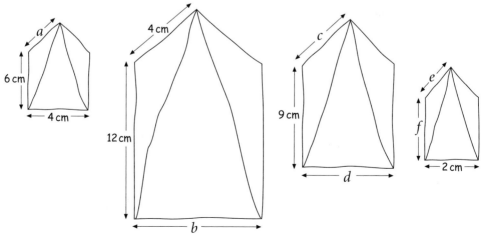

8 Which shapes are enlargements of which? Give the scale factors.

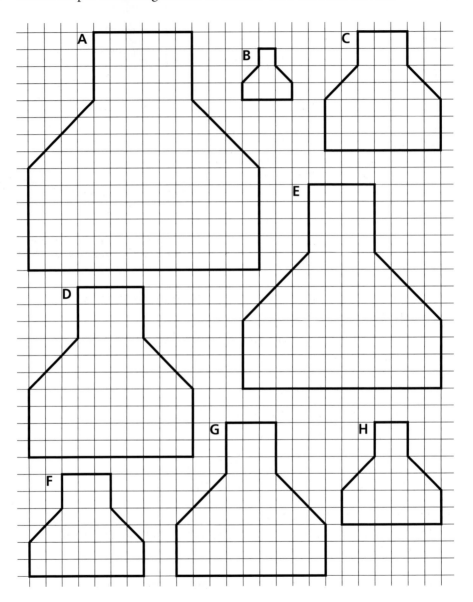

Section D

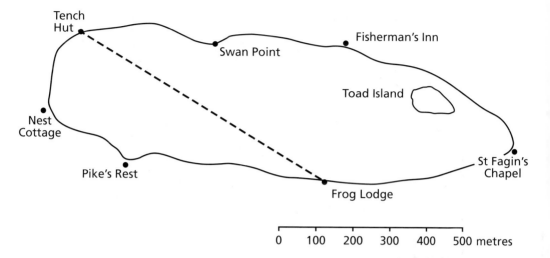

1 (a) Measure the distance in centimetres on the map between
       Tench Hut and Frog Lodge.

   (b) What is the actual distance, in metres, between Tench Hut
       and Frog Lodge?

2 Find these actual distances, in metres.

   (a) From Pike's Rest to Fisherman's Inn

   (b) From Nest Cottage to Swan Point

   (c) From Nest Cottage to St Fagin's Chapel

   (d) From Fisherman's Inn to Frog Lodge

3 The edge of the lake is curved.
   Describe how you could estimate from the map the perimeter
   of the lake.

4 How long would a distance of 1.2 km appear on a map drawn
   to the scale of the one above?

5 This is a map of Toad Island, drawn to a larger scale.
1 cm represents 20 m.

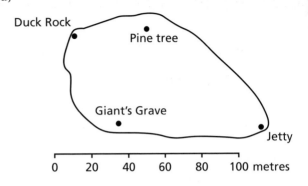

Find the actual distance, in metres, between each of these places.

(a) Duck Rock and the jetty

(b) Giant's Grave and the pine tree

(c) Giant's Grave and the jetty

6 How long would these appear, drawn to the scale of the map above?

(a) A fence 160 m long

(b) A road 2 km long

(c) A boat 16 m long

7 The map of Toad Island in question 5 is an enlargement of the map on page 75.

What is the scale factor of the enlargement?

Section E

1 Write each of these map scales as a ratio beginning 1 : ...

(a) 1 cm to 250 m   (b) 1 cm to 5 km   (c) 1 cm to 1 m

2 Write each of these scales as a ratio.

(a) 2 cm to 1 km   (b) 2 cm to 50 m   (c) 5 cm to 100 m

3 (a) Write the scale of the map on page 75 as a ratio.

(b) What is the scale of the map on this page as a ratio?

㉓ Approximation and estimation

Sections A, B and C

1 Work these out.

   (a) 40×2 (b) 30×60 (c) 6×40 (d) 500×5

   (e) 30×700 (f) 80×40 (g) 500×60 (h) 700×4

2 Work out a rough estimate for each of these.

   (a) 29×42 (b) 308×41 (c) 17×742 (d) 940×82

3 These five numbers go where the question marks are in this multiplication grid.

   `13` `27` `36` `73` `88`

   Use estimation to work out where they go.

×	?	?	49	?
18	234	648	882	1584
?	351	972	1323	2376
64	832	2304	3136	5632
?	949	2628	3577	6424

4 A farmer plants 28 rows of cabbages with 92 cabbages in each row. Roughly how many cabbages are there altogether?

5 A box contains 48 packets of pain-relief pills, with 32 pills in a packet. Roughly how many pills are there in a box?

6 Round each of these numbers to one significant figure.

   (a) 442   (b) 7902   (c) 52 088   (d) 97

7 Round each of these quantities to one significant figure.

   (a) 298 m   (b) 8 095 km   (c) 650 g   (d) 956 litres

8 Estimate each of these, roughly.

   (a) The area of a rectangular plot of land, 42 m by 293 m

   (b) The cost of having milk delivered for a year if the milk bill for a week is £5.80

   (c) The cost of vaccinating 780 pupils in a school with a vaccine costing £18 per pupil

Sections D, E and F

1 Work these out.

(a) 34×0.1 (b) 255×0.01 (c) 0.3×0.2 (d) 0.5×0.6

(e) 600×0.4 (f) 4.9×0.1 (g) 3.4×0.01 (h) 0.08×20

(i) 0.07×60 (j) 300×0.8 (k) 0.05×0.5 (l) 90×0.3

2 Round each of these numbers to one significant figure.

(a) 44.9   (b) 476   (c) 0.0426   (d) 5.721

(e) 7.0932   (f) 40.56   (g) 7.92   (h) 0.038

(i) 0.066   (j) 0.209   (k) 0.0062   (l) 1.0099

3 Write each of these quantities to one significant figure.

(a) 83.58 g   (b) 0.0654 kg   (c) 5.022 g   (d) 0.79 litre

(e) 0.3042 litre   (f) 0.0817 g   (g) 0.0061 g   (h) 475 kg

(i) 0.602 m   (j) 4.799 km   (k) 6467 km   (l) 51.8 m

4 A piece of fabric on a roll is 40 m long and 2 m wide.
Calculate its area.

5 Work out a rough estimate of the area of
each of these pieces of fabric.

(a) 48.2 m long and 2.9 m wide   (b) 19.2 m long and 2.15 m wide

(c) 27.6 m long and 1.05 m wide   (d) 32.6 m long and 1.85 m wide

6 For each of these,

     (i) Estimate the answer using numbers rounded to 1 s.f.

     (ii) If you can tell that your estimate is high or low, say so.

     (iii) Find the answer on a calculator and compare it with (i) and (ii).

(a) 320×2.3 (b) 470×793 (c) 0.401×9.1 (d) 128×0.03

(e) 0.81×676 (f) 0.064×620 (g) 5.5×912 (h) 550×1.4

(i) 0.0285×4.8 (j) 0.76×0.039 (k) 7.94×870 (l) 46×0.023

24 Bearings

Section A

1 Draw the rectangle ABCD to a scale of 1 cm to 4 km.

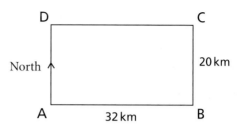

(a) What is the bearing of B from A?

(b) What is the bearing of C from A?

(c) What is the bearing of D from A?

(d) Find and label the point K whose bearing from A is 065° and from B is 325°.

(e) What is the bearing of K from D?

(f) How far is it from K to D?

2 This sketch map shows the route of a sailing race. It is in the shape of an equilateral triangle.

AB is on a bearing of 070°.

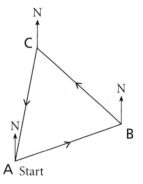

(a) Calculate the bearing of BC.

(b) Calculate the bearing of CA.

3 This diagram shows a square ABCD. B is on a bearing of 060° from A.

What is the bearing of

(a) C from A

(b) D from A

(c) B from E

(d) A from E

(e) D from C

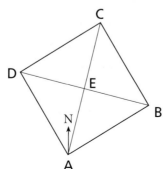

㉕ Using equations

Section A

1 Copy each wall. Work out the value of *x* for each wall.
Then find all the missing numbers in each wall.

(a)
```
      21
    ┌──┬──┐
  ┌──┼──┼──┐
  │8 │ x│ 3│
  └──┴──┴──┘
```

(b)
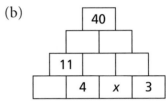

2 Copy each wall.
Work out the value of the unknown letter in each wall.
Then find all the missing numbers in each wall.

(a)
```
         63
      ┌────┬────┐
   ┌──┼────┼────┐
   │15│    │    │
┌──┼──┼────┼────┐
│ 9│  │  r │  6 │
└──┴──┴────┴────┘
```

(b)
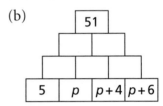

3 In these walls, pick your own letter to stand for one of the missing numbers. Work out all the missing numbers in each wall.

(a)
```
          43
       ┌────┬────┐
    ┌──┼────┼────┐
 ┌──┼──┼────┼────┐
 │3 │  │  4 │  7 │
 └──┴──┴────┴────┘
```

(b)
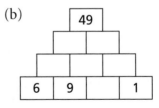

4 Work out all the missing numbers in each wall.

(a)
```
          54
       ┌────┬────┐
    ┌──┼────┼────┐
 ┌──┼──┼────┼────┐
 │  │ 5│    │  7 │
 └──┴──┴────┴────┘
```
This number... is the same as this number.

(b)
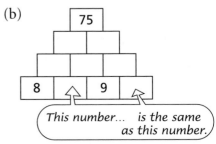

Section B

Show all your working and check that each answer works.

1 Solve each of these equations.

(a) $6a + 9 = 2a + 13$ (b) $2x + 18 = 3x + 6$ (c) $7y + 11 = 5y + 18$

(d) $8p + 15 = 5p + 27$ (e) $t + 16 = 3t + 10$ (f) $5x + 7 = 3x + 18$

2 (a) Multiply out the brackets from $5(n + 7)$.

(b) Use your answer to help you solve the equation $5(n + 7) = 2n + 44$.

3 Solve each of these equations.

(a) $3(x + 4) = 5x + 8$ (b) $4y + 37 = 7(1 + y)$ (c) $z + 26 = 5(z + 2)$

(d) $10(g + 8) = 8g + 85$ (e) $3(6 + h) = 7(h + 2)$ (f) $4(p + 1\frac{1}{2}) = p + 9$

Section C

1 Turn each number puzzle into an equation and solve it.

(a) Charlie thinks of a number.
He multiplies it by 7.
He adds 15.
His answer is 10 times the
number he started with.

What number did he start with?

(b) Pam thinks of a number.
She adds 8 to the number.
She multiplies the result by 4.
Her answer is 6 times the
number she started with.

What number did she start with?

2 Karl and Liz both think of the same number.
Karl multiplies his number by 4 and adds 9.
Liz adds 5 and multiplies the result by 3.
They both get the same answer.

What was their starting number?

3 Jackie and Angela both think of the same number.
Angela adds 7 and then multiplies by 4.
Jackie adds 2 and then multiplies by 8.
They both get the same answer.

What number did they both start with?

Section D

1 Solve $6z - 3 = 4z + 5$. (Add 3 to both sides to start with.)
Check your solution works.

2 Solve these equations.

(a) $7b - 4 = 4b + 11$ (b) $3t + 7 = 5t - 11$ (c) $79 + 3d = 11d - 1$

3 Solve these equations.

(a) $4y - 3 = 8y - 7$ (b) $7x - 3 = 10x - 9$ (c) $5h - 6 = 4h - 1$

4 Solve these equations.

(a) $3(x - 7) = 5x - 39$ (b) $5x + 9 = 9(x - 3)$ (c) $7(t - 3) = 2t + 9$

(d) $3(y - 8) = 9(y - 10)$ (e) $3(2x - 5) = x + 20$ (f) $2(y + 12) = 7y + 9$

5 Tom and Jo each think of the same number.
Tom says: 'I take off 13 and then multiply by 6.'
Jo says: 'I multiply by 2 and then add 14.'
They are surprised to find that they both end up
with the same number.

Work out what number they were both thinking of at first.

Section E

1 Nick and Ewan have the same number of sweets.

Nick has 4 bags full of sweets
and 8 sweets left over.

Ewan has 3 bags full of sweets
and 23 left over.

If each bag holds the same number of sweets, how many sweets are
in each bag?

2 Classes 9G and 9H go on a school trip.
Unfortunately the coach does not arrive to take them home.
The teacher organises some taxis but only 5 are available immediately.

9G have 4 full taxis and 4 pupils have to wait.
9H have 1 full taxi and 22 pupils have to wait.

There are the same number of pupils in each class and
each taxi holds the same number of pupils.

How many pupils are in each taxi?

3 John and Sue collect some apples.
John fills 2 bags but finds that 6 apples in the bags are rotten.
Sue fills 4 bags but finds that 14 apples are rotten.
Each bag holds the same number of apples.

If they both end up with the same number of good apples,
how many apples does each bag hold?

4 For the school fair Sam brings 15 bags of homemade fudge.
On the way to school Sam and his friends eat 3 pieces of
fudge from each bag.

Suppose there are f pieces of fudge in each bag.

(a) Which expression tells you the total number of pieces of fudge left
after Sam and his friends had eaten some?

$$\boxed{15f - 3} \qquad \boxed{15(f - 3)} \qquad \boxed{f - 45}$$

(b) The number of pieces left is the same as the number in 12 full bags.

How many pieces of fudge are in a full bag?

5 Alicia and Matthew are the same age.
Alicia multiplies her age by 2 and adds 6.
Matthew multiplies his age by 5 and takes off 33.
Each result is the same as the age of their maths teacher.

Work out the ages of the teacher, Alicia and Matthew.

㉖ Distributions

Section A

1 Here are the weights in grams of a litter of baby mice.

    (a) Find the median weight.

    (b) Find the range of the weights.

> 19, 18, 16, 20, 17, 16, 19

2 These are the tail lengths in mm of another litter of mice.
Find the median and range.

> 46, 58, 57, 48, 55, 51, 49

3 Zoe keeps lots of mice.
Here are her records for the numbers of mice born in each litter.

6, 5, 7, 5, 4, 5, 6, 7, 6, 7, 5, 8

    (a) What was the mode for the number of baby mice?

    (b) What was the median number of baby mice?

4 Sam also keeps lots of mice.
These are the records of the number of mice born in each litter over the last few months.

Number of mice in litter	Tally	Frequency
4	||	2
5	＋＋＋＋	5
6	＋＋＋＋ |	6
7	＋＋＋＋ |||	8
8	|||	3
9	|	1

    (a) How many litters of mice are there?

    (b) What is the range of the number of mice in a litter?

    (c) What is the mode for the litter size?

    (d) What is the median litter size?

5 Look at your results for questions 3 and 4.
Use them to compare litter sizes for Zoe's mice and Sam's mice.

Section B

1 Here is a stem-and-leaf table showing the length of time, in minutes, that it took some pupils to solve a mathematical puzzle.

0	4 6 8
1	0 2 3 5 5 7
2	1 3 4 6 8 8 9 9
3	0 1 2 2 5
4	2 6 7 9
5	5

(a) Find the range of the times.

(b) Find the number of people who completed the puzzle.

(c) Find the median time.

(d) Find the modal group.

2 In a biology experiment 25 pupils were growing sunflowers in two different types of soil.
Here are the heights of their plants in cm after one month.

Soil A

| 26, 30, 48, 59, 62, 25, 34, 36, 41, |
| 43, 47, 56, 51, 45, 24, 32, 38, 32, |
| 41, 40, 57, 60, 21, 39, 47 |

Soil B

| 29, 33, 44, 48, 51, 53, 35, 42, 50, |
| 46, 48, 26, 29, 33, 51, 30, 28, 41, |
| 47, 54, 50, 42, 35, 46, 48 |

(a) For each type of soil

    (i)   make a stem-and-leaf table

    (ii) find the range of heights

    (iii) find the median height

    (iv) find the modal group

(b) Use these results to compare the two soils.

3 24 pupils weighed their school bags. Here are the weights in kg.

2.4	3.1	1.9	2.8	3.4	4.2	4.0	4.4
3.1	2.6	2.7	3.9	1.8	2.0	1.5	3.4
4.3	2.6	2.9	3.2	2.2	3.7	4.1	4.2

(a) Make a stem-and-leaf table to show these weights.

(b) What is the range of the weights?

(c) What is the median weight?

Section C

1 Here are the weights, in kg, of some pupils' school bags.
   Calculate the mean weight.

   2.5   3.6   3.4   4.1   3.7   2.8   2.9   3.0

2 Anita kept a record of how much she spent on presents for her family.
   Here are the amounts.

   £2.99   £1.75   £4.50   £3.10   £2.75   £1.80   £3.60   £4.99   £3.49

   Calculate the range of the amounts and the mean amount,
   to the nearest penny.

3 These are the times, in seconds, for the competitors in
   two heats of a 400 metre race.

   Heat 1:   46.25   47.10   46.09   46.32   46.17
   Heat 2:   47.84   45.96   46.28   46.47   48.15

   (a) Find the mean time and the range of the times for each heat.

   (b) Compare the results in the two heats.

4 Each pupil in a class counted how many words they had
   written on one side of a piece of A4 paper.
   Here is the data they collected.

   Girls   304 328 118 247 306 120 218 125 272 228 279 335
   Boys   348 206 116 259 212 186 315 107 164 319 243 96
           137 311

   (a) Find the mean and range for the girls and boys separately.

   (b) Write two sentences comparing the number of words
       for girls and boys.

   (c) Find the mean number of words for the whole class.

5 Pete has been given the results of five of his examinations:

   48%   62%   59%   71%   68%

   (a) What is his mean score?

   (b) He now receives his last result, 59%. What is his new mean score?

Section D

1 Sasha conducted a survey in her class to find out how many pens,
   pencils and other writing implements each pupil had.
   Her sister Hadeel did a similar survey in her class.
   Here are their results.

Sasha's class

4	2	8	7	9	12	15	7	5	4	3	6	4
2	1	12	14	17	15	7	18	1	9			

Hadeel's class

12	15	4	7	9	10	14	16	2	3	7	10	8
12	17	15	16	20	9	4	15	17	19			

(a) Make a grouped frequency table for each class.
    Use groups 1–5, 6–10 and so on.

(b) Draw a frequency bar chart for each class.

(c) What is the modal group for Sasha's class?

(d) What is the modal group for Hadeel's class?

2 Here are the heights, in metres, of the pupils in a Year 9 class.

Boys   1.42   1.35   1.53   1.47   1.45   1.39   1.51   1.56   1.48
        1.36   1.28   1.51   1.30   1.34   1.47

Girls   1.35   1.55   1.40   1.29   1.32   1.52   1.41   1.26   1.35
        1.28   1.44   1.38   1.39   1.30   1.43

(a) Make grouped frequency tables for boys and girls.
    Use intervals 1.20–1.30, 1.30–1.40 etc.
    (Make sure you know which interval 1.30 and so on will go into.)

(b) Draw a frequency chart for boys and another for girls.

(c) Write a sentence or two comparing the two sets of heights.

3 This chart shows the ages of people
   using a library one day.

   (a) Which is the modal interval?

   (b) How many people under 20 used
       the library?

   (c) What can you say about the age
       of the oldest user of the library?

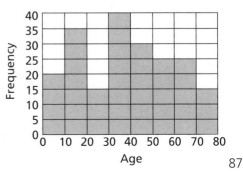

87

Section E

1 These are the ages of the people on a coach trip.

6	6	7	7	7	8	8	9	9	9	9	10
11	11	11	12	12	12	13	13	14	32	35	44

(a) What is the modal age of the people on the coach trip?

(b) What is their median age?

(c) Calculate the mean age.

(d) Which average, if any, do you think gives the best representation of the age of the people on the coach?
Give a reason for your answer.

2 The following donations were put into a collecting box by a party of visitors.

2p	5p	5p	10p	10p	10p	10p
10p	20p	20p	50p	50p	£1	£1
£1	£1	£5	£10			

(a) Find the mean, median and mode for the donations.

(b) Which, if any, do you think gives a typical value for a donation? Explain your answer.

3 The heights of some British and some Japanese students of the same age are to be compared.

British students – heights in cm			
154	145	142	156
149	152	150	138
145	158	139	140

Japanese students – heights in cm			
134	142	140	153
141	137	151	137
142	147	140	143

(a) Find the range of heights for the British students and for the Japanese students.

(b) Find the median and the mean for both sets of data.

(c) Write a sentence or two comparing the heights of the students.

27 A sense of proportion

Sections A and B

1 Which of these problems can be solved by multiplying by 3?

   A A steak recipe uses a tablespoon of peppercorns for two people.
         How many tablespoons will I need for steak for six people?

   B A mug weighs 250 grams.
         What will be the weight of three mugs like this one?

   C It takes 30 minutes to bake 2 apples.
         How long will it take to bake 6 apples?

   D Neeta takes half an hour to prune her roses.
         How long would it take three people to prune these roses?

   E Three ounces are equivalent to 85 grams.
         How many grams are equivalent to nine ounces?

2 Some of these problems involve direct proportion
   and some of them do not.
   Solve only those problems that involve direct proportion.

   A It took me about four hours to paint my bedroom
         How long would it take two people to do the same job?

   B I have a tin of paint that covers an area of $30\,m^2$.
         What area would 5 of these tins of paint cover?

   C One mile is equivalent to 1.6 km.
         How many kilometres are equivalent to 10 miles?

   D Hebe takes ten minutes to drive 2 miles in a busy city centre.
         How long will it take her to drive 20 miles along the motorway?

   E A recipe for meat sauce for 4 people uses 500 g of minced beef.
         How much minced beef would you need for meat sauce
         for 12 people?

3 One litre is equivalent to 1.76 pints.

(a) Which of these five graphs shows the link between litres and pints?

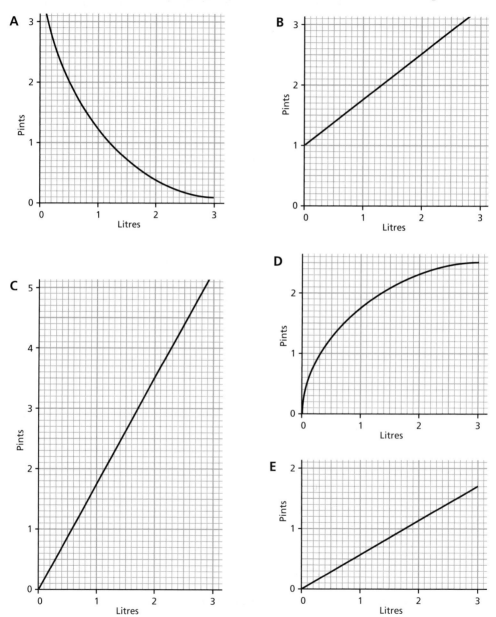

(b) Use the correct graph to estimate the number of pints in 2.4 litres.

Section C

This is a conversion graph (January 2002) for pounds sterling (£) and euros (€).

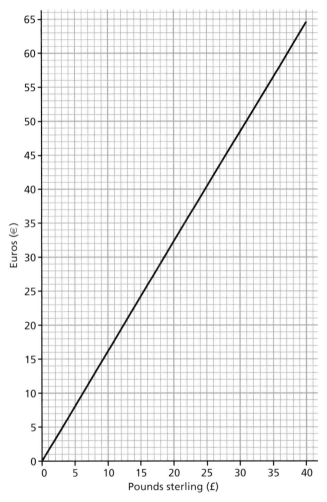

1 Use the graph to convert
   (a) £10 to euros
   (b) £39 to euros
   (c) €45 to pounds sterling
   (d) €42 to pounds sterling

2 (a) Convert £60 to euros. Show your method clearly.
   (b) Convert €75 to pounds sterling. Show your method clearly.

3 Floor areas are usually measured in square metres or square yards.

30 square yards is equivalent to 25 square metres.

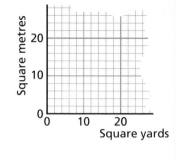

(a) Draw a conversion graph for converting areas up to 60 square yards.

(b) Use the graph to convert (to the nearest whole number)

  (i) 10 square yards to square metres

  (ii) 40 square metres to square yards

(c) Which of these rules links the number of square metres (m) to the number of square yards (y)?

$$m = \frac{y}{30} \times 25$$

$$m = \frac{y}{25} \times 30$$

Section D

1 To make fish pie for 4 people I need 60 grams of butter.
How much butter will I need to make fish pie for 3 people?

2 A garden centre uses 30 kg of compost to fill 6 identical tubs.
How much compost would they use to fill 10 of these tubs?

3 5 packets of crisps cost £1.50.
How much will 2 packets of these crisps cost?

4 15 copies of a booklet weigh 2.75 kg.
How much will 90 of these booklets weigh?

5 The total length of 3 fencing panels is 5.80 m.

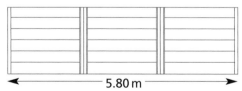

5.80 m

What is the total length of 8 fencing panels, correct to two decimal places?

6 Fiona changed £30 for 3933 Japanese yen.
How much would she get for £55?

28 Fractions

Section A

1 Which of these fractions are equivalent to $\frac{1}{4}$?

$$\frac{5}{10}, \quad \frac{3}{12}, \quad \frac{8}{16}, \quad \frac{5}{20}, \quad \frac{7}{28}, \quad \frac{4}{40}$$

2 Which statement matches this diagram?

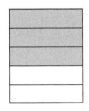 =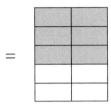

$$\frac{4}{5} = \frac{8}{10}$$
$$\frac{2}{3} = \frac{6}{9}$$
$$\frac{3}{5} = \frac{9}{15}$$
$$\frac{3}{5} = \frac{6}{10}$$

3 Draw a diagram to show $\frac{2}{5} = \frac{6}{15}$.

4 Which of these fractions are equivalent to $\frac{3}{5}$?

$$\frac{6}{8}, \quad \frac{9}{15}, \quad \frac{15}{25}, \quad \frac{25}{30}, \quad \frac{40}{50}$$

5 Simplify the following fractions as far as you can.

(a) $\frac{9}{18}$ (b) $\frac{14}{49}$ (c) $\frac{15}{18}$ (d) $\frac{25}{70}$

6 Write down three fractions equivalent to $\frac{1}{8}$.

7 Copy and complete these.

(a) $\frac{1}{5} = \frac{4}{\square}$ (b) $\frac{5}{8} = \frac{\square}{24}$ (c) $\frac{3}{8} = \frac{12}{\square}$ (d) $\frac{4}{9} = \frac{12}{\square}$

Section B

1 What fraction of each strip is shaded?

(a)

(b)

(c)

(d)

2 Work these out.

(a) $\frac{2}{3} - \frac{1}{5}$ (b) $\frac{3}{4} - \frac{5}{12}$ (c) $\frac{3}{4} - \frac{7}{10}$ (d) $\frac{3}{8} + \frac{1}{3}$

3 Find the missing fractions for these strips.

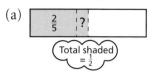

4 Write down a fraction larger than $\frac{1}{8}$ but smaller than $\frac{1}{3}$.

5 Put these fractions in order starting with the smallest.

$$\frac{1}{3}, \quad \frac{1}{4}, \quad \frac{1}{5}, \quad \frac{3}{5}, \quad \frac{5}{9}$$

Section C

1 Work these out as mixed numbers, simplified where possible.

(a) $2 \times \frac{4}{5}$ (b) $4 \times \frac{3}{8}$ (c) $9 \times \frac{5}{6}$ (d) $14 \times \frac{3}{4}$

2 Draw a diagram that shows $\frac{1}{4}$ of $7 = 1\frac{3}{4}$.

3 Work these out and give your answers as mixed numbers.

(a) $\frac{1}{8}$ of 13   (b) $\frac{1}{2}$ of 15   (c) $\frac{1}{3}$ of 8   (d) $\frac{1}{4}$ of 13

4 Give your answers to these as mixed numbers in their simplest form.

(a) $\frac{3}{4}$ of 22   (b) $\frac{5}{6}$ of 10   (c) $\frac{3}{8}$ of 20   (d) $\frac{2}{9}$ of 24

5 4 people share 11 cakes equally. What does each person get?

6 Find the missing number in each calculation.

(a) $\frac{1}{3}$ of $\square = 3\frac{1}{3}$ (b) $\frac{1}{4}$ of $\square = 4\frac{1}{2}$ (c) $\frac{1}{5}$ of $\square = 2\frac{4}{5}$

Section D

1 Work these out.

(a) $5 \div \frac{1}{2}$ (b) $5 \div \frac{1}{5}$ (c) $9 \div \frac{3}{4}$ (d) $6 \div \frac{2}{3}$

29 Constructions

Section A

1 For each of these, draw the triangle accurately, label the vertices with their letters, then measure and record the missing lengths and angles.

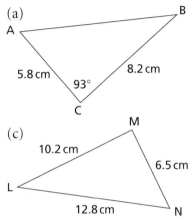

(a)

(b)

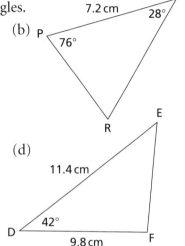

(c)

(d)

Section B

1 Draw a line *l* and mark a point A on it.
Use ruler and compasses to construct a line through A perpendicular to *l*.

Use ruler and compasses to bisect one of the right angles.
(This is the construction for an angle of 45°.)

2 Draw a line *l* and mark a point A on it.

Draw an arc, centred at A, cutting the line *l* at B.

Do not alter the radius.
Draw an arc, centred at B, to cross the first arc at C.

(a) What is the size of angle CAB?

(b) Bisect angle CAB using ruler and compasses.

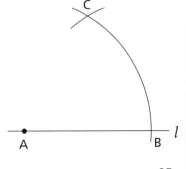

30 Division

Do not use a calculator for this work

Section A

1 Work these out.

(a) $\dfrac{45}{15}$ (b) $\dfrac{56}{14}$ (c) $\dfrac{120}{15}$ (d) $\dfrac{225}{75}$

(e) $\dfrac{120}{24}$ (f) $\dfrac{150}{25}$ (g) $\dfrac{144}{16}$ (h) $\dfrac{105}{15}$

2 Work these out.

(a) $\dfrac{360}{60}$ (b) $\dfrac{5500}{50}$ (c) $\dfrac{14\,000}{20}$ (d) $\dfrac{4800}{800}$

(e) $\dfrac{63\,000}{90}$ (f) $\dfrac{2700}{300}$ (g) $\dfrac{560}{70}$ (h) $\dfrac{2800}{400}$

3 Work these out.

(a) $\dfrac{420}{14}$ (b) $\dfrac{3000}{75}$ (c) $\dfrac{720}{18}$ (d) $\dfrac{9000}{150}$

(e) $\dfrac{1250}{25}$ (f) $\dfrac{1280}{16}$ (g) $\dfrac{15\,000}{25}$ (h) $\dfrac{6000}{240}$

Section B

1 Work these out as decimals.

(a) $\dfrac{14}{20}$ (b) $\dfrac{12}{40}$ (c) $\dfrac{4}{50}$ (d) $\dfrac{34}{170}$

(e) $\dfrac{18}{300}$ (f) $\dfrac{45}{900}$ (g) $\dfrac{28}{700}$ (h) $\dfrac{9}{200}$

2 Work these out as decimals.

(a) $\dfrac{0.8}{20}$ (b) $\dfrac{2.4}{40}$ (c) $\dfrac{1.8}{20}$ (d) $\dfrac{2.1}{700}$

(e) $\dfrac{1.4}{200}$ (f) $\dfrac{0.6}{30}$ (g) $\dfrac{0.9}{18}$ (h) $\dfrac{6.4}{80}$

3 Work these out as decimals, making clear which are recurring decimals.
$\dfrac{1}{6}, \dfrac{2}{6}, \dfrac{3}{6}, \dfrac{4}{6}, \dfrac{5}{6}$

Section C

1 Work these out.

(a) $\dfrac{8}{0.4}$ (b) $\dfrac{21}{0.3}$ (c) $\dfrac{48}{0.6}$ (d) $\dfrac{80}{0.2}$

(e) $\dfrac{4.5}{0.5}$ (f) $\dfrac{0.28}{0.7}$ (g) $\dfrac{0.08}{0.4}$ (h) $\dfrac{160}{0.8}$

2 Work these out.

(a) $\dfrac{9}{0.03}$ (b) $\dfrac{2.5}{0.05}$ (c) $\dfrac{24}{0.04}$ (d) $\dfrac{0.3}{0.06}$

(e) $\dfrac{4}{0.8}$ (f) $\dfrac{50}{0.02}$ (g) $\dfrac{20}{0.05}$ (h) $\dfrac{5.6}{0.07}$

3 How many glasses of capacity 0.06 litre can be filled from a bottle holding 0.9 litre?

4 How many flooring blocks 0.2 m long can be cut from a piece of wood 3.4 m long (ignoring the amount wasted in cutting)?

Section D

1 Work out a rough estimate for each of these divisions.

(a) $\dfrac{82}{1088}$ (b) $\dfrac{394}{22}$ (c) $\dfrac{6983}{51}$ (d) $\dfrac{588}{19}$

(e) $\dfrac{3107}{48}$ (f) $\dfrac{8214}{385}$ (g) $\dfrac{38\,416}{763}$ (h) $\dfrac{62}{1025}$

2 Work out a rough estimate for each of these divisions.

(a) $\dfrac{78.4}{5.2}$ (b) $\dfrac{119}{2.8}$ (c) $\dfrac{1578}{42.1}$ (d) $\dfrac{8.22}{0.39}$

(e) $\dfrac{61.6}{0.52}$ (f) $\dfrac{0.062}{2.1}$ (g) $\dfrac{18.8}{0.022}$ (h) $\dfrac{58.3}{0.31}$

3 Estimate how many curtain lengths, each 3.18 m long, can be cut from a roll of fabric 59.5 m long.

4 Estimate how many chairs 0.48 m wide will make a row 19 m long.

31 Indices

Section A

1 Copy and complete these factor trees.

(a)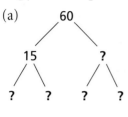

(b)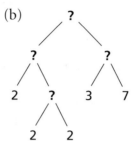

(c)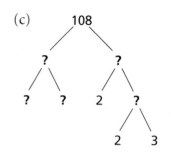

2 (a) Make a factor tree for 72.

(b) Copy and complete the statement below to write 72 as a product of prime numbers.

72 = ☐ × ☐ × ☐ × ☐ × ☐

3 Find the missing prime number in each of these products.

(a) $45 = \blacksquare \times 3 \times 5$

(b) $75 = 3 \times \blacksquare \times 5$

(c) $56 = 2 \times 2 \times \blacksquare \times 7$

(d) $84 = 2 \times 2 \times \blacksquare \times 7$

4 (a) Write 120 as a product of primes.

(b) Use your product to decide which of these numbers are factors of 120.

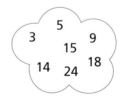

*5 (a) Write 156 as a product of primes.

(b) How does your product show that 12 is a factor of 156?

Section B

Do not use a calculator for questions 1 to 4.

1 Which of the expressions below is equivalent to 'four to the power five'?

A $\boxed{4 + 4 + 4 + 4 + 4}$ B $\boxed{4 \times 4 \times 4 \times 4 \times 4}$

C $\boxed{5 \times 5 \times 5 \times 5}$ D $\boxed{4 \times 5}$

2 Write these products using indices.
(a) $2 \times 2 \times 2 \times 2 \times 2$ (b) $5 \times 5 \times 5 \times 5$
(c) $3 \times 3 \times 3 \times 3 \times 3 \times 3$

3 Find the value of each of these.
(a) 2^4 (b) 3^3 (c) 5^2 (d) 4^3

4 (a) Make a factor tree for 144.
(b) Write 144 as a product of prime numbers using indices.

5 Use a calculator to find the value of each of these.
(a) 2^8 (b) 3^6 (c) 4^7 (d) 7^4

6 Find the value of each of these.
(a) $2^3 + 3^2$ (b) $4^2 \times 3^3$ (c) $7^2 + 5^3$ (d) $3^6 \div 3^2$

7 Write the following numbers as the products of primes using indices.
(a) 27   (b) 96   (c) 196   (d) 900

8 What is the missing number in each statement?
(a) $270 = 2 \times 5 \times 3^\blacksquare$ (b) $500 = 2^2 \times 5^\blacksquare$
(c) $396 = 2^2 \times 3^\blacksquare \times 11$ (d) $675 = \blacksquare^3 \times 5^2$

Section C

1 Write these numbers   (i) in figures
                          (ii) as powers of 10

   (a) one thousand   (b) ten thousand

   (c) one million   (d) one hundred million

2 Work out the value of each of these.

   (a) 6×10^3 (b) 27×10^4 (c) 40×10^6 (d) 95×10^5

3 Work out the value of each of these.

   (a) 2.4×10^3 (b) 1.75×10^4 (c) 26.2×10^2 (d) 8.62×10^5

4 Work out the value of each of these.

   (a) $620 \div 10^3$ (b) $151 \div 10^2$ (c) $4.26 \div 10^3$ (d) $90.2 \div 10^2$

5 In October 1964, the Japanese opened their high speed Bullet Train line
   from Tokyo to Osaka.

   (a) The Series 0 Bullet Train could travel at 210 km per hour.
       How many metres per hour is this?

   (b) The power of the Bullet Train was 11.8 megawatts.
       How many watts is this?

   (c) The Bullet Train on display in the National Railway Museum in York
       travelled a total distance of 10.2 gigametres while in service.
       How many metres is this?

6 Match the equivalent pairs of numbers.
   Which one is the odd one out?

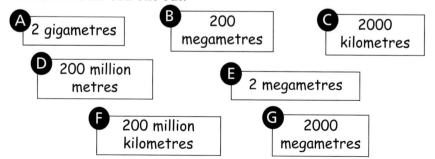

A 2 gigametres
B 200 megametres
C 2000 kilometres
D 200 million metres
E 2 megametres
F 200 million kilometres
G 2000 megametres

Section D

1 Write each of these as a single power of 4.

   (a) $4^2 \times 4^3$ (b) $4^4 \times 4$ (c) $4^3 \times 4^4$ (d) 4×4^5

2 What is the missing number in each statement.

   (a) $3^{\blacksquare} \times 3^4 = 3^6$ (b) $6^2 \times 6^{\blacksquare} = 6^9$ (c) $8^{\blacksquare} \times 8 = 8^6$

3 Write each of these as a single power of 5.

   (a) $5^4 \div 5^2$ (b) $5^3 \div 5$ (c) $5^7 \div 5^2$ (d) $5^8 \div 5^7$

4 Write each of these as a single power of 3, where possible.

   (a) $3^3 \div 3$ (b) $3^2 + 3^3$ (c) $3^4 \times 3^3$ (d) $3^6 - 3^3$

5 Simplify these expressions.

   (a) $b^2 \times b^4$ (b) $x^3 \times x^5$ (c) $y \times y^4$ (d) $r^2 \times r^5$

6 Simplify these expressions.

   (a) $t^3 \div t$ (b) $m^5 \div m^3$ (c) $n^6 \div n^2$ (d) $z^8 \div z^5$

7 In each diagram, the expression in each square is found by multiplying the expressions in the circles on either side of it.

   Copy and complete each diagram.

   (a) 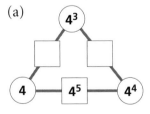   (b) 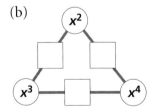   (c)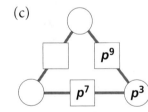

Mixed questions 4

1   What are the scale factors
    of these enlargements?

    (a) A to B

    (b) B to C

    (c) C to A

    (d) A to D

    (e) D to B

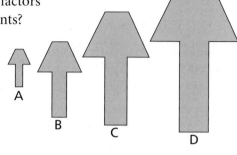

2   (a) Work out $2^2 \times 3 \times 5^2$.

    (b) Write 450 as a product of primes.

3   The diagrams show a design for a new
    logo and an enlargement of it.
    The diagrams are not drawn to scale.

    (a) What is the scale factor of the enlargement?

    (b) Work out the missing measurements, a and b.

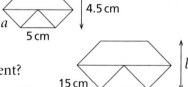

4   Solve these equations.

    (a) $5r - 8 = 2(r - 1)$ (b) $4(a - 1) = 6(a - 3)$

    (c) $15z - 4 = 8z + 31$ (d) $3(2t + 1) = 8t - 15$

5   A map is drawn to a scale of 1 cm to 20 km.

    (a) A lake is 1.7 cm long on the map. How long is the real lake?

    (b) Two towns are 112 km apart. How far would this be on the map?

6   This is an addition wall.
    Work out the value of x.

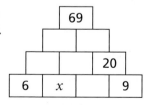

7   A recipe for 4 people needs 220 g of cheddar cheese.
    How much cheese will be needed for 7 people?

8 Ship P is 8 km north of lighthouse L. The boat B is on a bearing of 115° from ship P and is on a bearing of 055° from the lighthouse L.

(a) Draw an accurate diagram marking the position of boat B, using a scale of 1 cm to 1 km.

(b) How far is boat B from ship P?

(c) What is the bearing of the lighthouse from boat B?

9 (a) A ribbon is 30 m long and 0.06 m wide. What is its area?

(b) 6300 leaflets are to be delivered by 90 volunteers.
How many leaflets will each volunteer need to deliver if they each deliver the same number?

10 Here is a conversion graph for Australian dollars and pounds sterling.

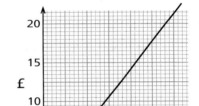

(a) Use the graph to convert

(i) £5 to A$

(ii) A$40 to pounds

(b) A video costs £12 in Britain and A$31 in Australia.

In which country is it cheaper?

11 Here is a stem-and-leaf table for the marks in a test.

```
1 | 2  3  3  5  8  9
2 | 1  1  3  5  5  7  8
3 | 0  2  4  4  7
```

(a) How many pupils took the test?

(b) What was the median mark?

(c) What is the range of the marks?

12 Estimate the answers to these calculations.

(a) 0.21×69 (b) 0.036×0.32 (c) $\dfrac{22}{0.38}$ (d) $\dfrac{0.064}{0.312}$

13 Simplify these expressions.

    (a) $c^3 \times c^5$ (b) $d^8 \div d^2$ (c) $e^4 \times e$

14 Here are the prices of some secondhand cars in two showrooms.

    Showroom A: £4.4k   £3.5k £8k   £5.3k £12k   £4.9k

    Showroom B: £7k   £6.5k £5.9k £8k   £6.8k £9k

    Find the mean price and the range of the prices for each showroom.
Write two sentences comparing the prices.

15 (a) Draw accurately triangle RST where RS = 8 cm,
       ST = 7 cm and RT = 5.5 cm.

    (b) Construct, using ruler and compasses only, the angle bisectors of
       angle R and angle S.

    (c) Label the point where the two angle bisectors meet as X.
       How far is X from the point T?

16 Write these map scales as ratios. Simplify the ratios if possible.

    (a) 1 cm to 50 m   (b) 4 cm to 100 m   (c) 1 cm to 5 km

17 Some pupils were given a mental maths test and a written maths test.
Here are their scores.

Mental maths	6	18	14	12	18	10	14	19	9	15
Written test	15	42	43	30	36	28	25	30	28	34

    Draw a scatter graph to show this information.

18 Work out   (a) $\frac{5}{8} - \frac{1}{4}$ (b) $\frac{1}{8}$ of 12   (c) $\frac{1}{2} + \frac{1}{6}$ (d) $15 \times \frac{1}{4}$

19 Marty and Sam are blowing up balloons for a party.
Marty blows up all the balloons from 3 packets each containing
n balloons, but 8 of his balloons burst.
Sam's packets of balloons each contain one more balloon than Marty's
packets, and Sam blows up two of these packets.

    They end up with the same number of inflated balloons.

    Write down an equation and solve it to find the number of
balloons in Marty's packets.